STO

ALLEN COUNTY PUBLIC LIBRARY

P9-AGS-493

# CONTEMPORARY ASTRONOMY

Edward J. Devinney, Jr.
*University of South Florida*

Haywood Smith, Jr.
*University of South Florida*

Sabatino Sofia
*University of South Florida*

520
W 49c

00016

80 1002 3

Charles E. Merrill Publishing Company
*A Bell & Howell Company*
Columbus, Ohio

Published by
Charles E. Merrill Publishing Company
*A Bell & Howell Company*
Columbus, Ohio 43216


This book was set in Times Roman and Serif Gothic.
The Production Editor was Susan Ziegler.
The cover was designed by Will Chenoweth.
Cover photograph: Crescent earthrise prior to TEI, lunar farside in foreground. Courtesy of NASA.
Chapter opening illustrations by Larry Hamill.

Copyright © 1975 by Bell & Howell Company. All rights reserved. No part of this book may be reproduced in any form, electronic or mechanical, including photocopy, recording, or any information storage and retrieval system, without permission in writing from the publisher.

International Standard Book Number: 0-675-08727-9

Library of Congress Card Catalog Number: 74-27269

1 2 3 4 5 6 7 8 9—80 79 78 77 76 75

Printed in the United States of America

1885457

# Preface

Several events which have taken place in the last two decades have led astronomy to what may be termed its golden age. Chief among these events is the space program, whose influence, either direct or indirect, appears in each new advance of this science.

In general terms, the advances may be classified in three different types. First, either new or better detectors were developed in each band of the electromagnetic spectrum. Second, the space program developed vehicles which took some of those detectors beyond the earth's atmosphere. Finally, all the data received could be reduced and interpreted only by virtue of the existence of the powerful computers which have only appeared in the last decade. The combined action of all these advances has expanded our understanding of the universe to limits never envisioned even in the recent past.

Living during this golden age of astronomy constitutes by itself a challenge. We may choose to totally ignore this science, in which case we miss an opportunity that may never return during our lifetime. We may, on the other hand, feel that it is our duty, in addition to a potential source of excitement and satisfaction, to share of the uncovering of the best guarded secrets of nature. In this case, however, it is necessary for us to acquire the background in astronomy which will allow us to understand these findings. In other words, we must learn the language, as well as the basic procedures, employed by astronomers.

Although such undertaking may at first sight appear rather difficult, most of the astronomical developments, while difficult to carry out in practice, are based on relatively simple principles. Moreover, they are carried out by strict adherence to the scientific method, which also is simple to understand if not to practice. It is easy to understand also the practical aspect of astronomy as a training ground for the use of the scientific method in its purest form.

The book is designed for a one-term course of introductory astronomy for nonscientists. With some additional material it could also cover a two-term course. Little mathematical or physical background is needed. The book is organized in three parts. The first part contains background information and instrumentation. The second part deals with the planetary system and the stars in both their observed and theo-retical properties. The third part covers our galaxy, external galaxies, and cosmology. We selected those topics which are relevant to a current understanding of the universe and its parts. The casualities of our selec-tion process were topics which, although interesting on their own, had little relevance to the understanding of our large-scale universe. For example, the many types of intrinsic variable stars are not dealt with in the book; the only variable stars covered are the cepheids, which are very important to distance determinations.

The topics that survived this process are discussed clearly, and the discussions are carried to the present, including not only what is known, but also what remains to be discovered. This conveys an essential flavor of working astronomy, as we feel the student of our book not only will be aware of some of the problems currently at the forefront of astronom-ical research, but will also understand why they are problems. Informa-tion gathered from observations is stressed throughout.

On completing this course, the student will be familiar with the current state of the field, including those areas where the search is still going on. At the same time, he will have the bases and the vocabulary which will allow him to follow the nonspecialized literature describing the advances to come. In other words, he will be able to fully share the excitement of discovery in a field whose aim is to understand nature itself in all its grandeur.

The authors express their thanks to William M. Protheroe and Stephen Hill for their suggestions and careful reading of the manuscript.

We deeply appreciate the help and enthusiasm of our wives during the book's creation.

# Credits

**Part 1—Opener** Courtesy National Radio Astronomy Observatory (NRAO)

**Chapter 1—Fig. 1–18** Courtesy of NASA **Fig. 1–19** Courtesy of NASA **Fig. 1–20** Courtesy of NASA **Fig. 1–21** From the *American Ephemeris and Nautical Almanac*

**Chapter 2—Fig. 2–5** By permission of Rare Book Division, New York Public Library

**Chapter 3—Fig. 3–1** Courtesy Hale Observatories **Fig. 3–2** Courtesy Hale Observatories **Fig. 3–14** Yerkes Observatory photograph **Fig. 3–20** Courtesy Hale Observatories

**Chapter 4—Fig. 4–2** Yerkes Observatory photograph **Fig. 4–5** Yerkes Observatory photograph **Fig. 4–9** (both) Kitt Peak National Observatory **Fig. 4–10** Courtesy Hale Observatories **Fig. 4–11** University of Hawaii, Institute for Astronomy **Fig. 4–12** Courtesy Hale Observatories **Fig. 4–16** (both) Courtesy of Perken-Elmer Corporation, Boller & Chivens Division **Fig. 4–20** Courtesy of NASA **Fig. 4–22** (both top photos) Courtesy of NASA **Fig. 4–22** (bottom photo) Courtesy of Harvey Tananbaum, Center for Astrophysics, Cambridge, Massachusetts **Fig. 4–25** Courtesy of National Astronomy and Ionospheric Center; the NAIC is operated by Cornell University under contract with the National Science Foundation **Fig. 4–26** Courtesy NRAO **Fig. 4–27** From James M. Moran, *Astrophysical Journal* 185 (Oct. 1973): 552, by permission of the author and publisher (University of Chicago Press) **Fig. 4–28** Courtesy NRAO **Fig. 4–31** Courtesy Richard M. Goldstein, Jet Propulsion Laboratory of the California Institute of Technology

**Chapter 5—Opener** Pioneer Plate courtesy of NASA **Fig. 5–1** (all) Courtesy of NASA **Fig. 5–2** (both) Courtesy of NASA **Fig. 5–3** Courtesy of NASA **Fig. 5–4** Courtesy of NASA **Fig. 5–5** (all) Courtesy of NASA **Fig. 5–6** Courtesy of NASA **Fig. 5–8** International Planetary Patrol Photograph **Fig. 5–9** (both) courtesy of NASA **Fig. 5–10** Courtesy of *Science* 183 (Jan. 1974), pp. 187–89, "The Clues About the Early Solar System" by W. D. Metz, copyright 1974 by the American Association for the Advancement of Science **Fig. 5–11** Courtesy of NASA **Fig. 5–12** Courtesy of NASA **Fig. 5–13** (both) Courtesy of NASA **Fig. 5–14** Courtesy of NASA **Fig. 5–15** (both) Courtesy of NASA **Fig. 5–16** Courtesy of NASA **Fig. 5–17** (both) Courtesy of NASA **Fig. 5–18** NASA photo by the Lunar and Planetary Laboratory, University of Arizona **Fig. 5–19** NASA photo by the Lunar and Planetary Laboratory, University of Arizona **Fig. 5–20** Courtesy Hale Observatories **Fig. 5–21** (both) Courtesy of NASA **Fig. 5–22** Courtesy of

NASA    **Fig. 5–23** Courtesy Hale Observatories    **Fig. 5–24** Courtesy Hale Observatories

**Chapter 6—Fig. 6–1** (photo) Courtesy Hale Observatories    **Fig. 6–2** Courtesy Hale Observatories    **Fig. 6–3** Yerkes Observatory photograph by Mr. E. E. Barnard    **Fig. 6–4** Courtesy Hale Observatories    **Fig. 6–5** From a figure by Dr. L. H. Aller, with permission    **Fig. 6–6** (both) Courtesy Hale Observatories    **Fig. 6–7** Courtesy of Dr. Jesse L. Greenstein, California Institute of Technology    **Fig. 6–8** Yerkes Observatory photograph    **Fig. 6–9** Courtesy Hale Observatories    **Fig. 6–10** Courtesy Hale Observatories    **Fig. 6–11** From Peter van de Kamp, *Handbuch der Physik,* Vol. 50, page 188 by permission of the author and the publisher, Springer-Verlag    **Fig. 6–12** Courtesy Hale Observatories    **Fig. 6–13** Courtesy Hale Observatories    **Fig. 6–15** From *The Binary Stars* by Robert G. Aitken, by permission of Dover Publications, Inc.

**Chapter 7—Fig. 7–6** (photo) Courtesy Hale Observatories    **Figs. 7–11** and **7–12** Data from *Astrophysical Quantities* 2nd Ed. by C. W. Allen (London: Athlone Press), by permission of the author.

**Chapter 8—Fig. 8–3** Data from *Astrophysical Quantities* 2nd Ed. by C. W. Allen (London: Athlone Press), by permission of the author    **Figs. 8–10, 8–12,** and **8–13** Courtesy of Robert E. Wilson, Goddard Institute for Space Studies

**Chapter 9—Fig. 9–3** Data from *Astrophysical Quantities* 2nd Ed. by C. W. Allen (London: Athlone Press), by permission of the author    **Fig. 9–4** From an article by J. B. Oke in the *Journal of the Royal Astronomical Society of Canada,* Vol. 49, by permission of the author and publisher.    **Fig. 9–5** Courtesy Hale Observatories    **Fig. 9–6** Courtesy Hale Observatories    **Fig. 9–7** Courtesy Hale Observatories    **Fig. 9–8** Courtesy of G. H. Herbig, Lick Observatory    **Fig. 9–11** Adapted, with permission, from "Stellar Evolution Within and Off the Main Sequence," *Annual Review of Astronomy and Astrophysics,* Vol. 5, page 585. Copyright © 1967 by Annual Reviews, Inc. All rights reserved.    **Fig. 9–13** Courtesy Hale Observatories    **Fig. 9–15** (both) Courtesy Hale Observatories    **Fig. 9–17** Courtesy Hale Observatories    **Fig. 9–21** Courtesy Hale Observatories

**Chapter 10—Fig. 10–1** Courtesy Hale Observatories    **Fig. 10–2** Courtesy Hale Observatories    **Fig. 10–5** Courtesy Hale Observatories    **Fig. 10–8** Courtesy J. Oort, F. Kerr, and G. Westerhout

**Chapter 11—Fig. 11–1** Courtesy Hale Observatories    **Fig. 11–3** Courtesy Hale Observatories    **Fig. 11–4** Courtesy Cerro Tololo Inter-American Observatory, operated under National Science Foundation Sponsorship, photograph by Victor M. Blanco    **Fig. 11–5** Courtesy Hale Observatories    **Fig. 11–6** Courtesy Hale Observatories    **Fig. 11–7** Courtesy Hale Observatories    **Fig. 11–8** Courtesy Hale Observatories    **Fig. 11–9** Courtesy Hale Observatories    **Fig. 11–10** Courtesy Hale Observatories    **Fig. 11–11** Courtesy of Harlan Smith, University of Texas at Austin, McDonald Observatory

**Color Plates**
**Plates 1, 2, 3, 4, 5,** and **6** Courtesy of NASA
**Plates 7, 8, 9, 10, 11, 12, 13, 14, 15, 16, 17,** and **18** Copyright by the California Institute of Technology and Carnegie Institution of Washington

# Contents

Part 1: Principles and Practices                                    1

**1  The Sky**                                                       3
Perceiving the Sky   3
Observable Effects of the Earth's Daily Rotation   5
Consequences of the Earth's Yearly Revolution   12
The Phases of the Moon   17
Eclipses of the Sun and Moon   20
The Stars   24
Questions   27

**2  The Beginnings of Modern Astronomy**                          31
Order from Chaos   31
Dynamical Astronomy   32
Laws of Planetary Motion   37
The Force of Gravitation   46
Consequences of the Gravity Law   47
Tidal Effects   55
Condensed Matter   59
Questions   60

**3  Matter, Light, and Stars**                                    63
Introduction   63
The Nature of Light   64
The Nature of Matter   72
The Spectrum of an Incandescent Object   83
Starlight and the Doppler Effect   86
Questions   89

**4**  Modern Observational Astronomy                    91

Introduction   91
The Telescope   92
Practical Considerations for Observing    110
Limitations to Telescopic Observation    117
Progress in Above-Atmosphere Observation    121
Ground-Based Astronomy in the Nonvisible Region    123
Questions   134


Part 2: Planets and Stars                              137

**5.**  The Solar System                                139

Introduction    139
The Earth and Moon    142
Earth's Companions    150
Lesser Bodies   169
Formation of the Solar System    176
Questions   178


**6**  Basic and Observable Properties of Stars         181

Introduction    181
Stellar Spectral Types    184
Binary Stars    190
Questions   197


**7**  Determining the Distances of Stars               199

The Importance of Stellar Distances    199
Direct Methods   199
Indirect Methods    206
Questions   218


**8**  Finding the Basic Properties of Stars            221

Introduction    221
Luminosities    221
Masses   224
Radii   228
Temperatures   235
Chemical Composition   237
Questions   241

9   Stellar Evolution                                              243
    Introduction   243
    The Hertzprung-Russell Diagram   244
    Nuclear Energy and Stellar Interiors   247
    Interstellar Medium, Nebulae, and Star Formation   251
    The Lives of the Stars   258
    Ashes to Ashes, Dust to Dust   271
    Questions   272


Part 3: Galactic Astronomy and Cosmology   277

10  The Milky Way Galaxy                                           279
    Introduction   279
    Star Counts   280
    Cluster Observations   283
    Interstellar Absorption   285
    Present View of the Galaxy   286
    Current Problems in Galactic Research   293
    Questions   294

11  The External Galaxies                                          297
    Introduction   297
    Classification of the Galaxies   299
    Determinaton of Galactic Distances   304
    Groupings of Galaxies   307
    Radio Studies of Galaxies   309
    Seyfert Galaxies   309
    Quasi-Stellar Objects   311
    Questions   315

12  Cosmology                                                      317
    Definitions   317
    The Cosmological Principle   318
    Cosmological Models   320
    The Three-Degree Blackbody Radiation   326
    Evolution of the Universe   328
    Questions   329

Appendix 1: Conversions and Data                                   331

Appendix 2: The Constellations                                     333

Index                                                              335

# 1

# PRINCIPLES AND PRACTICES

**1** The Sky

**2** The Beginnings of Modern Astronomy

**3** Matter, Light, and Stars

**4** Modern Observational Astronomy

# 1 The Sky

## Perceiving the Sky

The present progress of astronomy may be traced to the great interest with which people have always regarded the sky. In times long past, the mode of living fostered a closer relationship to nature, resulting in a broad familiarity with the sky phenomena visible to the unaided eye. Our ancestors were motivated to learn about the sky by necessity, as well as curiosity, because all agricultural pursuits are dependent on the seasons and the seasons are announced by circumstances visible in the sky. Today, we reckon the seasons by a paper calendar and need no longer consult the sky, but in intense curiosity to understand our whole environment continues to motivate us as it did our ancestors.

In this book we will unfold a remarkable picture of the universe—discovered through the use of large telescopes and other sophisticated devices. However, the universe begins to be revealed through phenomena which can be seen by the unaided eye; so, let us introduce ourselves to the universe by an examination of these phenomena. No special equipment will be needed, just a decision to notice things.

What could we discover if we decided to find out for ourselves what is happening in the sky? Very likely we would become familiar with the patterns stars form—the constellations—and eventually we would know the sky like a map of our own locality. Our attention would quickly be drawn to the moon. A few patient hours of observation leave no doubt that it moves against the background of the stars. Furthermore, the moon's general appearance changes markedly as the nights go by and it shows its phases. Curiously, we would note that the phases are related to the *time* of moonrise. (Are you surprised, for instance, at the revelation that the full moon always rises at sunset?) Eventually, after some months of observation, we would become very suspicious about the nature of five particular starlike objects, because they too move against the background of the "fixed" stars. These "wandering stars" visible to the unaided eye are the planets Mercury, Venus, Mars, Jupiter, and Saturn. Continued scrutiny of Mercury and Venus would show that they never range very far from the sun. Either or both planets may be seen rising in the east before the sun as a "morning star," or following the sun down in the western sky as an "evening star." During the day, therefore, they either trail or lead the sun across the sky. Venus, in fact, is visible to the un-aided eye in the daytime sky when not too close to the sun, if one knows just where to look for it. Interestingly, many ancient cultures did not recognize the same planet in its roles as a morning and evening star.

Careful observations of the planets Mars, Jupiter, and Saturn over an extended time period would reveal that these planets do not move *uniformly* through the stars. Once a year, each planet will cease its forward (direct) motion through the stars, go backwards (retrograde) for a time, stop again, and move forward (direct) once more.* (See Figure 1–1.)

Figure 1–1. *The retrograde motion of a planet against the stars. The arrow indicates the planet's direct motion. The episode takes place over the course of weeks, and can easily be followed with the unaided eye for Mars, Jupiter, and Saturn.*

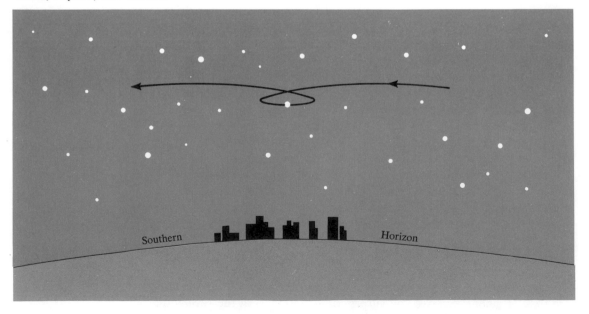

Southern                    Horizon

---

* Facing south, a direct motion is right-to-left, or eastward.

Finally, we cannot forget the most obvious celestial object, the sun. It also moves through the stars. That it does so may seem surprising, yet it is only necessary to observe the stars near the eastern horizon about one hour after sunset to see that the same stars appear higher in the sky as the weeks pass; thus, the sun must be changing its position with respect to them.

We are now aware of some of the many interesting and complex phenomena which can be observed in straightforward fashion, without the aid of a telescope. The complexity, however, is only apparent—a result of the earth's daily *rotation* on its axis and its yearly *revolution* about the sun.

## Observable Effects of the Earth's Daily Rotation

The result of the earth's daily rotation on its axis is the *apparent* rising of celestial objects in the east and their setting in the west. Actually, the detailed aspects of the apparent daily motion are a bit more complicated and *vary with the latitude of the observer.* To discuss these effects, it will be useful to think of the stars as glued to the surface of a very large sphere centered on the earth, referred to as the *celestial sphere.* The precise size of the celestial sphere is somewhat arbitrary, but "very large" ought to mean a radius for the sphere at least as large as the distance to the nearest star. The distance from earth to the nearest star is a few hundred thousand times the distance to the sun. As a consequence, our yearly journey about the sun does very little to change our perspective of the stars. (In figures in this text, the size of the earth will also be grossly exaggerated for clarity.)

Our experience, confirmed by Figure 1–2, indicates that an earth-bound observer can view only half of the celestial sphere at any moment, since the horizon divides the sky in two. The horizon of an average northern latitude observer, at 40° of latitude, is shown in Figure 1–3(a) as oriented on the earth, and in Figure 1–3(b) as it divides the sky. In both figures, the arrow points in the same direction as the axis of the earth's rotation. Since the celestial sphere is a great distance away, both the arrow and the axis seem to converge to the same point in the distance, like the familiar example of railroad tracks. This imaginary point on the sky about which the earth seems to rotate is the *north celestial pole.* It is approximately marked, by sheer coincidence, by the bright star Polaris. This star may be located by using the pointer stars of the Big Dipper, as shown in the star chart. While the earth rotates daily and carries our horizon with it, Polaris and a large region of the sky centered on the

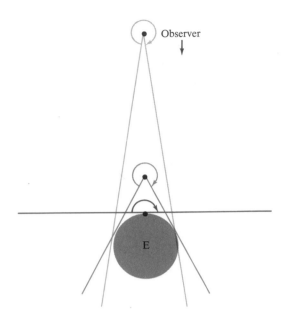

Figure 1–2. *The earth limits the observer's sky to one half of the complete celestial view, as the earth is approached. The arcs indicate the observer's field of vision.*

pole remain constantly above the northern horizon. Looking north, the stars seem to rotate counter-clockwise about the celestial pole. Figure 1–3(a) also shows an equal-sized cap of stars centered on the *south celestial pole* which never rises above our southern horizon. To see those stars, we must travel south. (This is the reason many northern observatories have southern stations.) Returning to the stars which do rise and set, Figure 1–3(b) indicates that the rising stars ascend on a slanting path, drifting southward as they rise. Their sloping path likewise causes them to drift northward as they set.

Traveling north or south over the earth causes dramatic changes in our perspective of the sky. Going northward, carrying our horizon with us, we find the north celestial pole rising higher in the sky until, on reaching the North Pole, we find the north celestial pole directly overhead. The cap of never-setting stars revolves in circles about the overhead point, or *zenith,* moving at all times parallel to the horizon. From this point, none of the stars of the south celestial hemisphere are visible. The celestial *equator,* which divides the celestial sphere in two, would be on our horizon.

Traveling southward from our 40° starting point causes new stars to begin showing over the southern horizon, and old ones to disappear beyond the northern horizon. On reaching the equator, we find that the north celestial pole has slipped all the way down to the north point on the horizon, and its cap of never-setting stars has shrunk to the vanishing point. On the southern horizon, directly south, we find the south celestial pole, the cap of never-rising stars having also vanished. The equatorial observer therefore has a special advantage: *all* stars are visible. Because

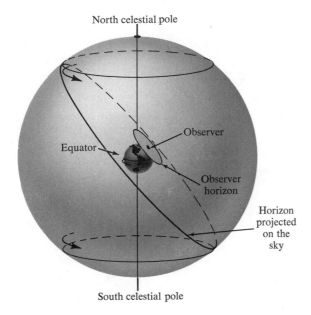

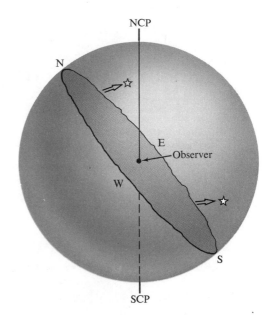

Figure 1–3. *(a) The body of the earth obscures half of the sky at any moment. Note the southern cap of stars which never rise, and the northern cap of stars which never set. (b) The observer's view of the sky. The earth's rotation causes the apparent rising and setting of celestial objects.*

of daylight, all will not be seen in one twenty-four-hour period, but they will be seen over the course of some months, as described below. At the equator, all stars rise and set. If we look directly east, we will see rising stars climbing straight up from the horizon, passing later overhead, and then setting straight down in the west. (See Figure 1–4.)

## About Time

The rotation of the earth causes the sky to sweep by once daily. The marking of the daily passage of objects in the sky thus forms the oldest and natural basis for reckoning time. We may use the sun, the stars, or even an imaginary point on the sky with a known position with reference to the stars as an hour hand. In addition to the object which moves, we must also have a reference line from which to judge the amount of the motion. Whatever this line is to be, it should be one which any observer could set up. How can this be done?

Consider yourself standing outside, and imagine a line originating at the north point of your horizon, which arcs overhead through the zenith,

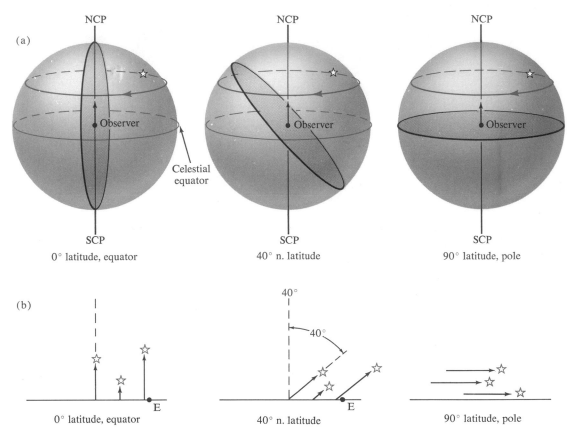

Figure 1–4. *(a) The observer's view of the sky for different latitudes. Note the very different appearance of the daily motion of the same stars. (b) View of the same group of stars from the different latitudes. They are seen rising, directly in the east, at latitudes of 0° (the equator) and 40° (a typical U.S. latitude). At the pole all stars are circumpolar; this group of stars skirts the horizon continually.*

and ends in the south point. This is your celestial meridian. The celestial meridian belongs to you—if you move east or west, it "follows" you. As the earth rotates, all sky objects drift westward across this line. The line is imaginary, of course, but we can set up instruments to locate it. After all, we can easily determine the location of the north, south, and zenith points (Fig. 1–5).

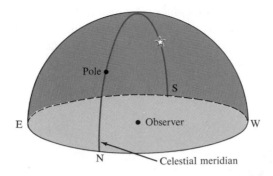

Figure 1–5. *The observer's celestial meridian, with a star at the instant of transit.*

Now that we have a reference line and objects which move, we can tell the time. As we will see below, it will make a considerable difference whether we choose the sun or the stars as timekeeper, since we already mentioned that the sun continually changes its position with reference to the stars. Thus, there are two types of "days": *sidereal* and *solar*. A day, or twenty-four hours, begins when the timekeeping object comes astride the meridian, and ends one complete circuit later.

### The Sidereal Day

Each 360° rotation of the earth results in one turn with reference to the stars. Each complete turn, therefore, brings the same stars to the celestial meridian of the observer. This time interval is the *sidereal* day.* Each sidereal day commences (00 hours, 00 minutes, 00.0 seconds) when a certain point on the sky crosses the meridian. It might be supposed that a bright star would make an ideal marker for this purpose. However, each star generally has its own very slow motion across the sky with respect to its neighbors; so, the stars are really not "fixed" points. Instead, a well-defined point with reference to the stars is used. The location of this point will be discussed later in the chapter. Each 15° of westward advance of the marker takes one hour of sidereal time, as 15° × 24 = 360°. Observatory clocks are set to run at the sidereal rate. This is an aid in finding the stars, as we shall see.

### Star Coordinates

The locations of points on the celestial sphere are specified in a manner which closely resembles that used for locating points on the spherical surface of the earth. On the earth, the location of any point may be

* Pronounced sī-der′-ee-al and means referring to the stars.

specified by giving its latitude and longitude, as shown in Figure 1–6. The system used for locating the stars on the celestial sphere is just a projection of the earth's system. Imagine the earth's rotation stopped, and a light at the center of earth projecting shadows of the earth's coordinates onto the

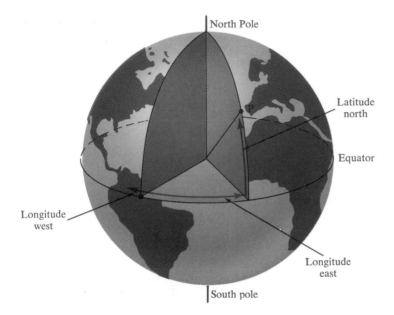

Figure 1–6. *Locating a point P on the spherical earth's surface using the two coordinates, latitude and longitude. O is the origin of the longitude coordinate.*

celestial sphere (Figure 1–7). On the sky, the latitude coordinate is called *declination,* while the longitude coordinate is known as *right ascension.* Just as latitude is measured in degrees north or south of the equator, declination is similarly measured with reference to the *celestial equator.* On the earth's surface, the longitude coordinate is measured in degrees (0–180°) east *or* west of the longitude line (the "prime meridian") which goes through Greenwich, England. Right ascension, on the other hand, is measured only eastward from a certain point on the celestial equator, the point where the sun is found on March 21. This is the same point which marks the beginning of the sidereal day. Because the stars form a clock, as we have noted, right ascension is divided into twenty-four hours. The time on a sidereal clock gives the right ascension of the sky which is then crossing the observer's celestial meridian.

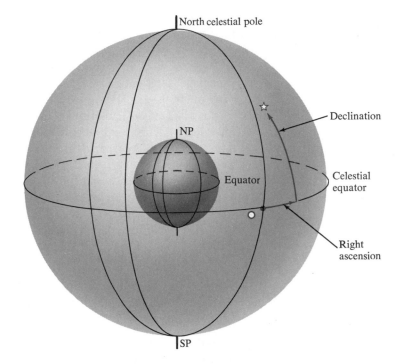

Figure 1–7. *Locating a point on the celestial sphere with the two coordinates, declination and right ascension. O is the origin of the right ascension coordinate. Note that, in contrast to longitude, right ascension is measured only eastward on the sky. (If your right thumb points to the north celestial pole, then your fingers curve in the eastward direction.) The figure depicts the earth stopped, and the earth coordinates projected onto the celestial sphere.*

## The Solar Day

If the sun is observed to rise among the stars, it may be seen to lag behind them each day by about four minutes. A solar clock, therefore, runs at a slower rate than a sidereal clock. If we were to set our solar clock running at a rate determined from observing the successive passages of the sun across our meridian, we would soon observe that most of the year the clock is not in step with the sun.

Why is the sun such an undependable timepiece? Actually, the fault does not lie with the sun, but with the earth. The sun's motion against the stars is only apparent. It is the result of the earth's motion. If, for example, we walk in a circle around a tree, we see it against a continually changing background—likewise with the yearly orbital journey of the earth about the sun. However, the earth does not move at a uniform rate. This is reflected as an apparent variation in the motion of the sun against the stars. A second, and more complicated, reason for the apparent variation is related to the fact that the earth's rotation axis is not perpendicular to the plane in which it journeys about the sun. A practical solar clock, therefore, completes twenty-four hours in the time it takes the sun, *on the average,* to make successive meridian crossings. This time is called *mean solar time* and it is the time in civil use.

The four-minute lag of the sun relative to the stars means the earth rotates an extra degree past one complete turn, so twenty-four hours of solar time requires a 361° rotation of the earth.

### Time Zones

Since the sun rises in the east, those east of us will see it sooner than us, and those west later. The clocks of those east of us should therefore be set ahead of ours, and those west, behind. In the days before rapid transportation, each town ran on its own time. To make an orderly transition in the change of time with longitude, the time zones have been devised. There are twenty-four of them around the globe, each representing a change of one hour. On continents, they have been considerably gerrymandered to agree with national, state, or city boundaries. In ocean areas, they jog around whole island groups. It is amusing to consider what happens when one travels completely around the globe, and consequently makes twenty-three time adjustments. When Magellan arrived in port after having sailed around the world, he was confused to find that his reckoning of the date differed by a day from the correct one. This, in fact, proved his claim to circumnavigation.

## Consequences of the Earth's Yearly Revolution

A complete journey of the earth about the sun requires about 365¼ days for 360° of revolution. Thus the daily motion of the earth in its orbit amounts to nearly one degree in orbit each day. We are not aware of this motion, but, as we noted earlier, this causes an apparent motion of the sun against the stars of one degree per day. Because of this motion, there is a continual change in the stars visible in the nighttime sky, as seen in Figure 1–8. The position of the earth in its orbit is related to the season, so that each season certain constellations are best seen. For example, when the earth has reached its place in orbit corresponding to winter, the constellation Orion will be predominantly visible in the night sky, opposite the earth from the sun. Six months later, Orion lies in roughly the same direction as the sun, and of course cannot be seen due to the brilliance of the daytime sky. At night the constellations *opposite* Orion are then in view.

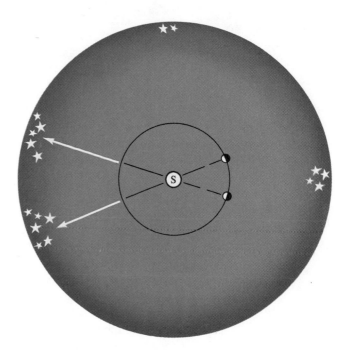

Figure 1–8. *The yearly apparent motion of the sun against the stars is just a reflection of the earth's annual motion. There is also a seasonal change in the stars visible in the night sky.*

### The Seasons

The seasons are the classical reminder of the passage of time. Each season has its own beauty and appeal. Why do we have seasons and not a monotony of climate? The seasons exist because the earth's rotational axis is

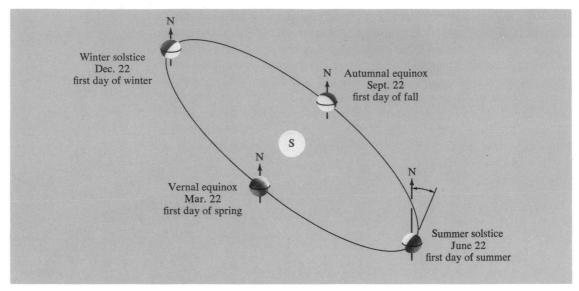

Figure 1–9. *The earth travels about the sun with its axis tilted by 23½°, causing a yearly variation in the angle with which sunlight strikes the earth. For example, the Northern Hemisphere receives the rays of sunlight more directly in summer than in winter. Note that the seasons in the Northern and Southern Hemisphere are reversed.*

tipped away from the perpendicular to the plane of the earth's orbit (Figure 1–9), by an angle of 23½°. Over the course of a year, the earth's axis maintains essentially the same direction in space. In Figure 1–9 we can see the daylit side of the earth for four important days of the year corresponding to the first day of summer, fall, winter, and spring. Note carefully how the Northern and Southern Hemispheres are situated with regard to receiving direct rays of the sun. As Figure 1–10 shows, the more directly the rays strike a portion of the earth, the more concentrated they are on a given area and the greater their heating effect. Figure 1–9 also shows that the seasons are reversed in the Northern and Southern Hemispheres. The most favorable date for receiving sunlight in the Northern Hemisphere occurs on June 22, when the earth's North Pole is most closely tipped towards the sun, while December 22 is the least favorable date for Northern Hemisphere dwellers. On the equinoxes, neither hemisphere is favored. This is not to say, however, that the hemispheres have similar climates on these days, because the earth takes some time to heat up and cool off. For example, the first day of spring is generally cooler than the first day of fall, since it is preceded by winter rather than summer.

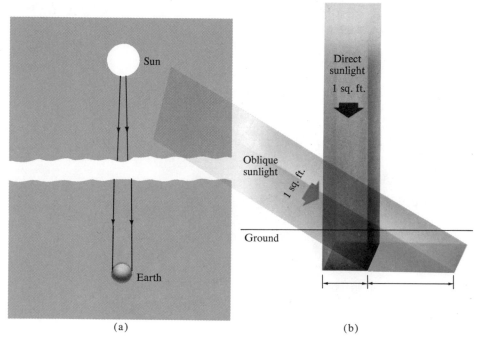

Figure 1–10. *(a) The earth, seen from the sun, subtends a very tiny angle, so that sunlight striking opposite sides of the earth left the sun traveling in practically the same direction. (b) The angle with which the rays strike the curved surface of the earth exerts a profound influence on the heat delivered. Each square foot of the earth receiving direct sunlight incurs the maximum heating effect. Oblique rays are distributed over a larger area and have less heating effect.*

### The Sun's Yearly Apparent Path on the Sky

We see also from Figure 1–9 that half the year the sun lies above (to the north of) the plane of the earth's equator and the other half it lies below. Since the celestial equator is a projection of the earth's equator, the sun's yearly path among the stars carries it above and below the celestial equator. Its path is shown in Figure 1–11, and is known as the *ecliptic*. By definition, the first day of spring occurs when the sun crosses the celestial equator on its northward journey. The intersection point is the point marking the beginning of the right ascension coordinate. The first day of spring is also the *vernal equinox* (vernal = spring, equinox = "equal night") because each point of the earth experiences twelve hours of sunlight and

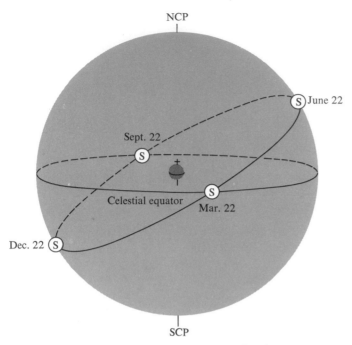

Figure 1–11. *The yearly apparent path of the sun, eastward on the celestial sphere with the earth at the center.*

twelve of darkness. On September 22, the sun crosses the celestial equator going southward. This date is the *autumnal equinox*. The sun's greatest northern and southern excursions from the celestial equator amount to $\pm 23\frac{1}{2}°$ and occur on June 22 and December 22. These dates are the *solstices* (= sun stand-still) because the sun has then reached the extreme of its travels and must stop and reverse.

An observer at a latitude of 40° finds that the daily journey of the sun in his sky on these four dates is as illustrated in Figure 1–12(a). Figure 1–12(b) shows the situation for an equatorial observer. (Why does an equatorial observer have only two different seasons?) It is left as an exercise for the student to explain the change in the daily appearance of the sun throughout the course of the year, as seen from the North Pole. (Hint: An observer at the North Pole has the celestial equator on the horizon.)

**The Length of the Year: The Calendar**

The approximately 365¼ days in which the earth completes an orbit of the sun (the year) are tabulated by a device known as the calendar. Everyone is aware that an extra calendar day must be added every fourth

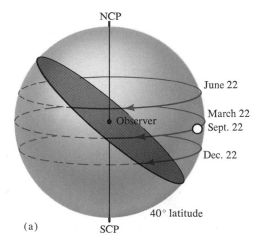

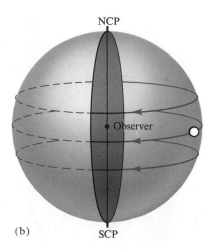

Figure 1–12. *(a) The changing aspect of the sun's daily motion with the season, for an observer at a latitude typical of the U.S. (b) The equatorial observer's seasonal change is less striking. The further from the equator, the more marked is the effect. Consider the effect at the poles for the most extreme example.*

year to account for the extra quarter day, but fewer people know what remedy is used to account for the fact that the year actually falls short of exactly 365¼ days by 11 minutes and 14 seconds. Occasionally, a leap year must be skipped. Only the century years evenly divisible by four hundred are leap years. The year 1900 was not a leap year, but 2000 will be.

## The Phases of the Moon

The *phase* of the moon refers to the appearance of its illuminated disk, as viewed by the earthbound observer. Every 29½ days, or one month, the moon goes through one cycle of its phases as it completes a circuit of the earth with reference to the sun. The cycle begins at the "new" phase, when the moon crosses the earth-sun line (Figure 1–13). The moon is then in the same general direction as the sun, so it rises with the sun and is up in the daytime sky. Its unilluminated hemisphere then faces the earth and, due to the brightness of the daytime sky, we are not aware of its presence. A slender crescent will be visible in a few days, as the moon's

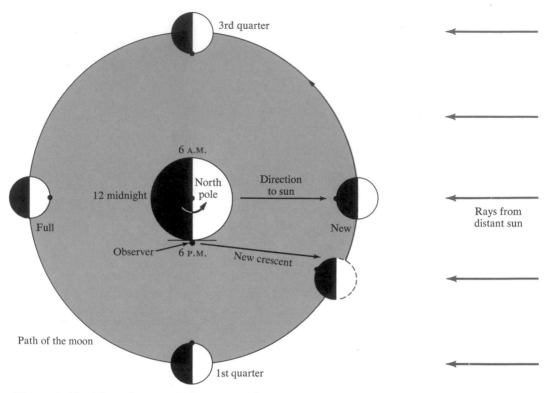

Figure 1–13. *The phases of the moon depend on its orbital position with respect to
the sun. Note that the observer sees the full
moon rise at sunset. From the figure, one
can estimate the time of moonrise and moonset for any phase. The dot indicates that the
moon rotates once on its axis during each
revolution about the earth.*

orbit carries it eastward away from the sun's direction. The best way to
see this crescent is to look in the western sky right after sunset. Just above
the western horizon, a silver sliver of a moon will be seen. At successive
sunsets, the moon will appear as a larger crescent, and be higher in the
sky (Figure 1–14). By the time the phase reaches full, about two weeks
later, the moon is rising in the east opposite the setting sun. Through the
next two weeks, the moon will continue to rise later and later, its illuminated disk thinning once again to a crescent as its rising time drifts closer
and closer to sunrise.

Figure 1–13 indicates how the phases depend on the position of
the moon in its orbit and how its rising and setting times are related to the

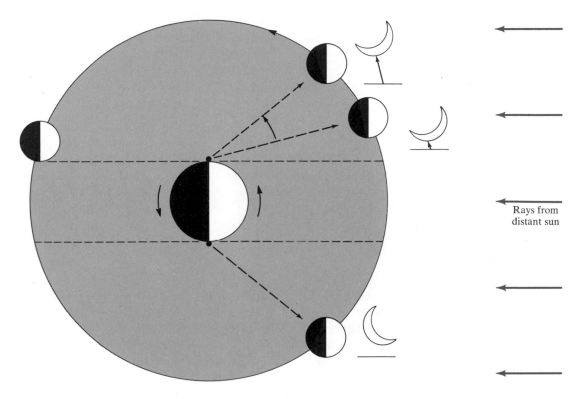

Figure 1–14. *The effect of a twenty-four hour interval on the appearance of the new crescent moon. The moon moves 13° in its orbit and appears higher in the sky over the western horizon. This continues until the moon has reached the eastern horizon and is rising while the sun is setting. Also indicated is the almost-new moon rising in the east just before the sun. How will it appear in twenty-four hours?*

Rays from distant sun

phases. Two other phases have received special names: the half moon, seen when the moon advances *one-quarter* of its orbit from new phase, is the "first quarter." The phase seen *three-quarters* of the orbit after new phase is likewise known as the "third quarter."

To ancient peoples, the 29½-day cycle of the phases was quite impressive. It held a kind of religious significance for many, and was further of a convenient length to serve as a calendar. Today, the lunar calendar is generally no longer used in the West, but its influence may still be seen in the lengths of the months. In fact, the word month itself derives from a word meaning moon. Incidentally, only half of the moon's surface is visible from the earth. During its journey around the earth it rotates on its axis so as to keep the same face pointed earthward.

## Eclipses of the Sun and Moon

Eclipses of the sun and moon are some of the most spectacular astronomical phenomena which can be witnessed. What conditions make them possible? We shall examine the appropriate circumstances for a solar eclipse first.

### Solar Eclipses

First of all, a solar eclipse can occur only during the new phase of the moon, because the moon must be in the sun's general direction as seen from a point on earth. However, a solar eclipse does not occur at every new moon, but only for about one new moon out of six. This is because the orbit of the moon does not lie in the plane of the earth's orbit, but is inclined by a 6° angle. As Figure 1–15 shows, the new moon is often sufficiently far away from the plane of the earth's orbit so that its shadow misses the earth entirely. However, the figure shows the moon must cross the earth's orbital plane twice each year, at places called *nodes*. The line LL′ connecting the nodes (the nodal line) tends to maintain the same general direction in space over the course of a year. Figure 1–15 indicates that about twice a year the new moon occurs "dangerously" close to a node and a solar eclipse is the result. Figure 1–16 shows the distinction between a total and a partial eclipse of the sun. The sun is completely hidden from an observer in the *umbra,* the darkest shadow. Viewers in the *penumbra* can see a partial eclipse. The conical umbra trailing the moon is cut by the earth's surface nearly at the point, so the diameter of the shadow is only about 150 miles (250 kilometers) at most. The shadow moves over the earth's surface at about 1,000 mph (1,600 kph). From Figure 1–17, it appears that the moon's motion overtakes that of a point on the earth so the shadow moves from west to east.

What is the experience of an eclipse like? In a region much larger than the ribbonlike path of totality, observers see a fraction of the sun covered, a partial solar eclipse (Figure 1–18). To observers only a few hundred miles away from the path of totality, the decrease in sunlight is not obvious, even though the majority of the sun's disk may be hidden. The experience in the path of totality, however, is completely stunning and must be personally witnessed to be fully appreciated. As the last bit

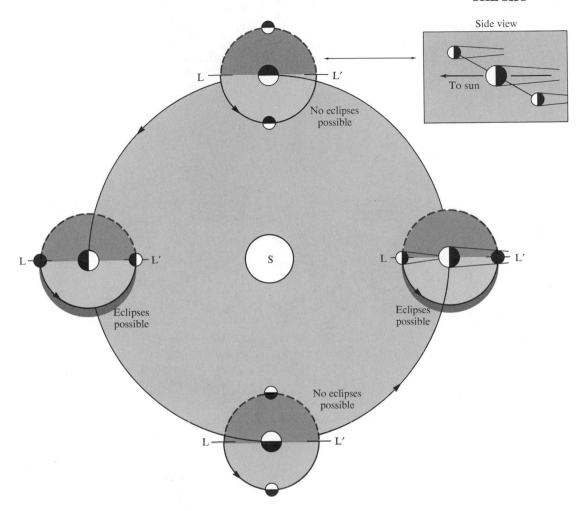

Figure 1–15. *When the new or full moon crosses the plane of the earth's orbit, eclipses are possible. LL', the line of nodes, maintains approximately the same direction in space over the course of one year.*

of the dazzling crescent of the sun is cut by the moon, it breaks up into "molten" drops of sunlight known as "Bailey's beads" (Figure 1–19) which ooze behind the moon. Then the sun immediately vanishes and a darkness like twilight engulfs the site. The brightest stars become visible, and usually Mercury and Venus can be seen. Around the sun is the delicate white *corona* (Figure 1–20), shining with the pale brightness of the full moon. Time stands still, and one can savor the rarefied feeling

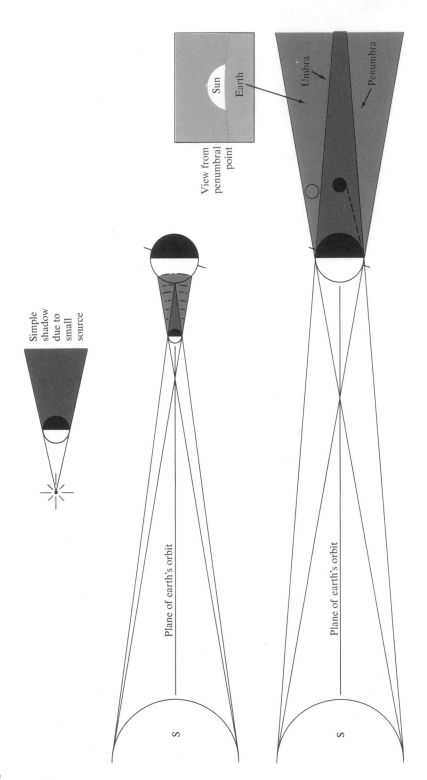

Figure 1–16. *Complex shadows due to large illuminating source. In the penumbra, a portion of the source can be seen. Note the relatively small region on earth which is in the moon's umbra. Note, by contrast, that eclipses of the moon are visible to all on the night side of the earth. The umbra of the earth is not completely dark due to "leakage" of light into it, light "refracted," or bent, by the earth's atmosphere.*

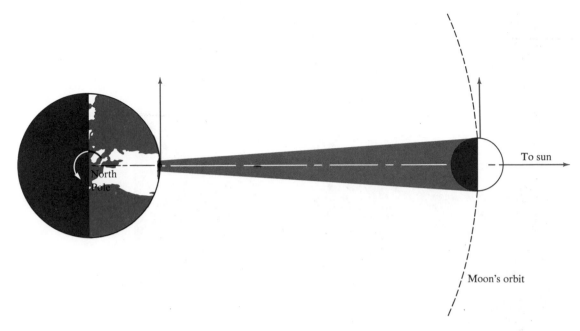

Figure 1–17. *The moon moves eastward faster than a point on the earth's surface; consequently, the shadow on earth is carried from west to east over the ground. The shadow moves at approximately 1000 miles per hour.*

because the longest possible interval of totality is near eight minutes and typically it is much less. If the chance arises to see a total solar eclipse, it must not be missed. The apparent rarity of this event is often somewhat misunderstood. There must be two solar eclipses per year, although there can be as many as five. *Each spot* on earth, however, experiences a total eclipse only once in about 450 years due to the very small area of the umbral shadow (see Figure 1–21). A list of solar eclipses is given in Table 1–1.

In some instances the point of the umbral cone does not quite reach the earth, because the moon's distance from the earth actually varies. The moon is then farther from the earth than usual, and appears smaller, so it cannot quite cover the sun. The eclipses are not total. An annulus of the sun surrounds the moon, giving rise to an *annular eclipse.*

## Lunar Eclipses

The full moon preceding or following a solar eclipse also occurs near a node, and thus the earth comes directly between the sun and the moon,

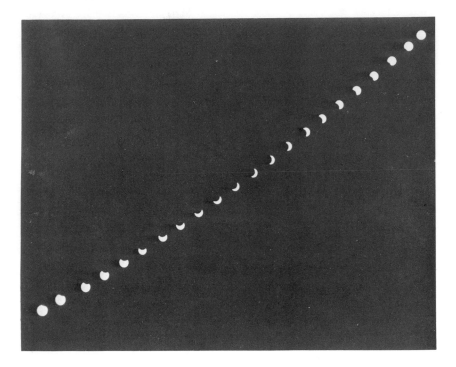

Figure 1–18. *The solar eclipse of 10 July 1972, photographed from a point in the penumbra, at the Goddard Space Flight Center, Greenbelt, Maryland.*

allowing the moon to enter the earth's shadow (Figure 1–15). Very little noticeable change occurs unless the moon reaches the umbra (the part of the shadow where the sun appears completely hidden). Instead of vanishing as expected, however, the moon usually appears bathed with a dull coppery light. It is sunlight scattered by the earth's atmosphere in the moon's direction. Lunar eclipses are beautiful and easy to observe because a lunar eclipse is visible to all on the night side of the earth.

## The Stars

Let us mention a few words about the background of stars which serves as the scene for all the action described in this chapter. The sky chart enables us to locate and identify the brighter unaided-eye stars. Each star may be found in the sky from its coordinates of right ascension

Figure 1–19. *The diamond ring, the last vestige of Bailey's beads. The inner corona is visible in this photo of the 7 March 1970 solar eclipse. The total eclipse was visible on the eastern coast of the United States.*

Figure 1–20. *The solar corona at mid-eclipse, 7 March 1970. Note the fringelike streamers which suggest that the sun has a global magnetic field, like that of a simple bar magnet, lying in this case in a direction from lower left to upper right. The corona waxes and wanes with an eleven-year cycle of solar activity; it appears somewhat different at each eclipse.*

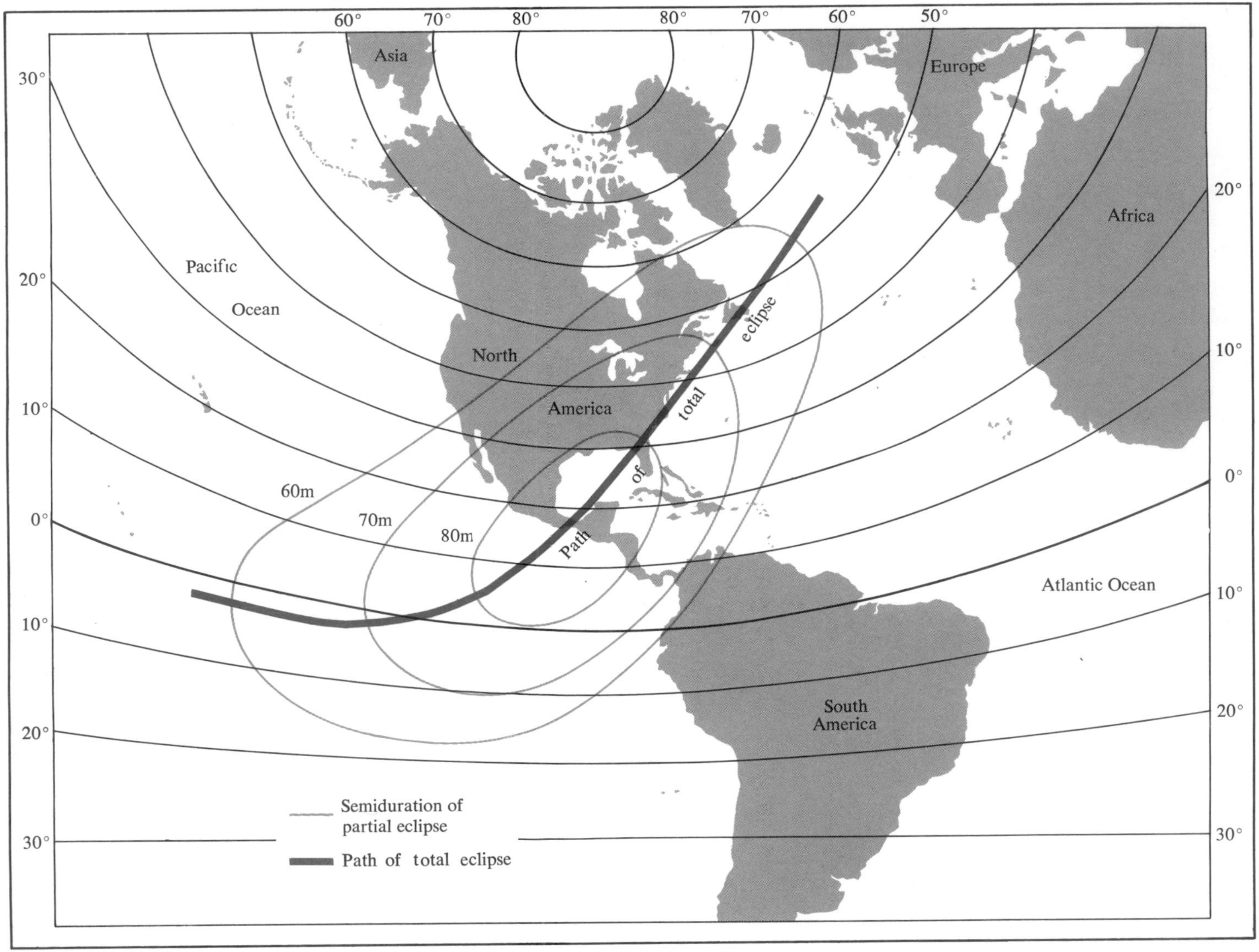

Figure 1–21. Plot showing the narrow path of totality of the 7 March 1970 eclipse, and the much wider region of visibility of the partial eclipse.

## TABLE 1–1

*Longer Solar Eclipses of the Relatively Near Future*

| Date | Length of complete eclipse (minutes) | Region of visibility |
|------|------|------|
| 23 Oct.  1976 | 4.8 | East Africa, Indian Ocean, Australia |
| 12 Oct.  1977 | 2.8 | NW South America |
| 26 Feb.  1979 | 2.7 | NW United States, Canada, Hudson's Bay |
| 16 Feb.  1980 | 4.3 | Central Africa, Southern India |
| 31 July  1981 | 2.2 | Central USSR, Siberia, N Pacific |
| 11 June  1983 | 5.4 | Indonesia |
| 22 Nov.  1984 | 2.1 | Indonesia, South America |
| 18 Mar.  1988 | 4.0 | Phillipines, Indonesia |
| 22 July  1990 | 2.6 | Northern Scandanavia, Arctic |

and declination. A help in locating a star is to note its approximate apparent brightness, known as its *magnitude*. The use of the term magnitude dates from ancient times. The Greek astronomer Hipparchus, in 150 B.C., divided the visible stars by apparent brightness into six categories. The brightest stars were categorized as "stars of the first magnitude," the second brightest were categorized as "stars of the second magnitude," etc. The stars of the sixth magnitude are those just visible to the average person on a clear, moonless night. The characteristics of stars will be discussed further in later portions of this book.

## Questions

1. How might the north circumpolar stars be used as a clock?

2. At what place(s) on the earth does the celestial equator pass directly overhead?

3. What time does the first quarter moon rise over China? Over the United States? At the North Pole (be careful!)?

4. Neither the earth's orbit about the sun nor the moon's orbit about the earth are circular. What relative distances conspire to give the longest possible duration for a total solar eclipse? The shortest?

5.  What is the sun's right ascension on March 22? On September 22?

6.  What is the most northernly and southernly declination of the sun (Hint: What is the tilt of the earth's axis?)

7.  Why could sidereal time not be practical for everyday use?

8.  Why are lunar eclipses so much more easily noted than solar eclipses?

# 2 The Beginnings of Modern Astronomy

## Order from Chaos

  The beginnings of modern astronomy can be set at about the fourteenth century. This was a time marked by an explosive growth of scientific inquiry, now referred to as the scientific revolution. For astronomy, these developments brought about some understanding of the nature of the solar system, specifically, the revelation that simple laws governed the apparently complex motions of its bodies. Thus the study of the motions of celestial objects (now known as *dynamical astronomy*) began. For a very long time, even until the turn of this century, the study of motions remained the principal field of inquiry for those calling themselves astronomers.

  The material in this chapter demonstrates that this early astronomy offered some important lessons to all science. Henceforth, there would be some reluctance to assume as correct even that which seemed obvious. *Neither opinions nor feelings regarding scientific matters were trusted unless they were tested successfully against observations of nature itself.* It further became clear that the job of science was not to engage in a discussion of the "ultimate truth" of the hypotheses needed to explain a phenomenon, but rather to uncover the simplest set of hypotheses which does explain a phenomenon. The most outstanding lesson, perhaps, was one uniquely astronomical: mere mortals could successfully pry loose even heavenly secrets.

## Dynamical Astronomy

### The Ptolemaic System

Well before the scientific revolution and the understanding of dynamics, curiosity aroused by the motions of the sun, moon, and planets against the fixed-star background begged some attempt at explanation. As noted in Chapter 1, these motions sometimes appear quite complex. At the beginning of the scientific revolution, a venerable explanation for these motions already existed in the so-called *Ptolemaic system.* The name derives from Claudius Ptolemaeus (Ptolemy), an Alexandrian Greek. By 150 A.D., he had melded the ideas of several previous investigators with his own into a coherent explanation for the various celestial wanderings. The Ptolemaic system was perhaps the inevitable result of marrying a thorough familiarity with these motions to the assumption that the earth was *an absolutely stationary center* for all of the astronomical activity observed. In the ancient theory, it was further assumed that all motions occurring had to be uniform in speed and circular in shape. The latter idea was the result of the ancient Greeks' regard for the circle as the perfect figure, and presumably the only one worthy of such noble sky objects.

The principal workings of the Ptolemaic system can be shown with the aid of the somewhat modern analogy of Figure 2–1. Here the earth is represented by the *nonrotating* center spindle of a record player, while the stars may be thought of as pasted on the record. Our VLP (very long playing) record rotates *westward* once every sidereal day: 23 hours, 56 minutes ($23^h 56^m$). All other objects revolve daily, but also have some motion of their own *eastward on the record.* The moon, for instance, may be thought of as a bug slowly making its way around a circular path on the record and completing one revolution in one month, about 29½ days. The sun likewise journeys once around its course on the record in about 365¼ days. These circular paths are known as *deferents.*

So far, Ptolemy's model accounts for the daily rising and setting of the sun, moon, and stars, and gives a reasonable explanation for the changing moon and sun positions with respect to the stars. An addition to the model, however, was needed to account for the motion of a planet such as Mars, whose eastward progress becomes interrupted once a year by a retrograde loop (as in Figure 1–1). Figure 2–1 shows that this looping can be mimicked by the addition of a smaller circle to the deferent. The small circle (an *epicycle*) carries the planet, and its center travels on the deferent. The grosser features of the motions of Mars, Jupiter, and Saturn can be reproduced well by this addition. Mercury and Venus must be

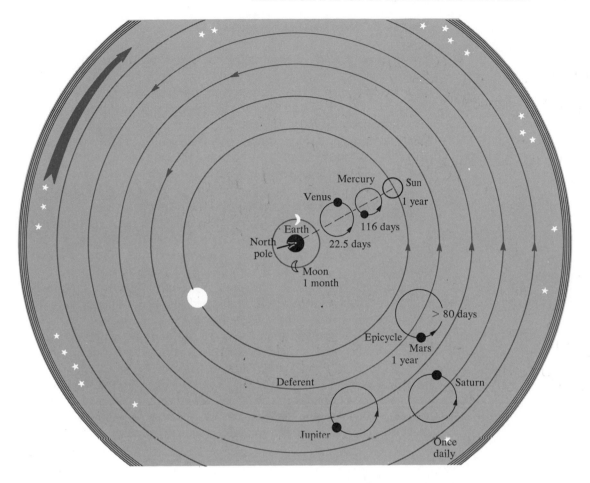

Figure 2–1. *The construction of the heavens, according to Ptolemy. The entire diagram revolves once daily around the nonrotating earth, while the objects themselves move independently and slowly in the opposite direction, like insects crawling around a rotating record. (This is a polar view.)*

treated differently. They are never seen far from the sun in the sky, so it was natural to consider that the centers of their circles (also called epicycles) had to be constrained on an imaginary line joining the sun to earth. In the Ptolemaic system, these two planets never travel behind the sun. As we shall see, this very important feature of the system was to fail its first observational test, although this was not to occur until 1610.

### The Sun-Centered Idea of Copernicus

By the sixteenth century, the original model of Ptolemy had become quite complicated. As observers accumulated more accurate data on planetary motions, they found irregularities which the simple epicycle and deferent was not able to reproduce. Thus the astronomers of the time found themselves adding epicycles to the epicycles, and introducing other tricks as well. The need for a simpler model was very evident to Nicholas Copernicus (1473–1543). Copernicus held that the sun, not the earth, was the center of all motions; thus he was opposing a *geo*centric view with a *helio*-centric one. His sun-centered hypothesis could not strictly be proved. On the other hand, Copernicus emphasized that it might be needlessly preju-dicial to *assume* that only the earth could mark the center. Furthermore, his heliocentric model was simpler, while explaining the observations equally well. Even among his contemporaries, this notion of greater simplicity had to strike a fundamental chord. Copernicus dispensed with the epicycles which had been required to produce the retrograde motions of Mars, Jupiter, and Saturn. He explained that planets closer to the sun circle it more quickly than those more distant, and claimed that these three planets were more distant from the sun than the earth. Retrograde motion, then, occurred whenever the earth overtook one of these (Figure 2–2). Contrary to some popular opinion, Copernicus's system did not con-sist solely of a sun-centered set of circular orbits. The motions of the planets, even in Copernicus's day, were observed to have small irregulari-ties, in addition to the obvious retrograde loops. Copernicus actually imagined the planets on small epicycles, set on heliocentric *circles*.

### A Scale Model of the Solar System

Copernicus found a most remarkable feature in his new system—it was possible to represent the solar system orbits in a scale model. This cannot be done for the Ptolemaic system. Incidentally, the method by which Copernicus determined the relative sizes of the planetary orbits serves as a prime example of how brainpower compensates for what might be assumed as a fatal deficiency—in this case, the inability to leave our earthbound observatory. (Another excellent example is the determination of stellar distances in Chapter 7.) Figure 2–3 shows the geometry for an inner planet P, such as Mercury or Venus. From the earth E we can measure the angle *a,* which eventually reaches a maximum value as the two planets orbit the sun. Geometry shows that our line of sight to P then lies tangent to its orbit, and makes a right angle with line SP. *All* triangles drawn with the observed angle *a* and a right angle must have the same porportions; thus, using an ordinary ruler, we can find the ratio SP/SE, the relative distances

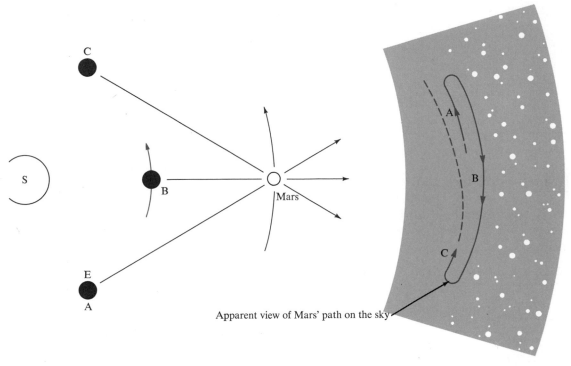

Apparent view of Mars' path on the sky

*Figure 2–2. Copernicus's view of the retrograde looping of Mars, or any planet more distant from the Sun than Earth. At A or C, our motion is principally toward or away from the planet, and thus we perceive its forward motion against the stars. At B, however, Mars seems to drift backwards.*

of the two objects from the sun. From Figure 2–3 it is clear that we cannot observe *a* for an outer planet (because then *a* would be the unmeasurable angle between the sun and the earth as seen from the planet), but it is possible to find angle *b*. First, it is a trivial matter to find out how long it takes *b* to go through 360°. Suppose we observe Mars at some patricular time to be opposite the sun (it is seen overhead at midnight). Record the date, and then patiently wait many months until this circumstance occurs again. During such a time interval, angle *b* has gone through 360°. This takes roughly two years, so angle *b* increases at about ½° each day. Now starting from an alignment, we can count off the number of days until Mars is seen at 90° from the sun (Figure 2–4), and find angle *b* by multiplying this number of days by the rate of ½° per day. Once again we have a triangle with two angles known, and all such triangles are proportional,

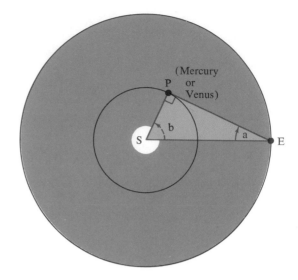

Figure 2–3. *Distance of Mercury or Venus from the Sun, compared to Earth-Sun distance, is determined from the shape of the triangle SPE. (See text for further explanation.)*

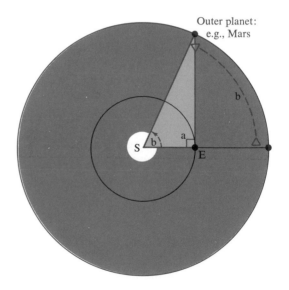

Figure 2–4. *The analog of Figure 2–3 for a planet more distant from Earth. The relative distance SM, compared to SE, is obtained by inferring angle b. (See text.)*

giving the ratio SE/SM. In this way one completes a scale model of the solar system. The only detail lacking is the actual scale for the distances involved. The earth-sun distance, the crucial element, was not to be revealed for some time; and, it might be added, after some grief.*

---

* One old method involved observing the rather rare crossings of Venus over the disc of the sun from different positions on earth. Thus the French astronomer la Caille set out for India in 1761, but his ship was delayed in a skirmish and he arrived a few weeks too late for the event. He decided to *wait* the six years until the next event. On the appointed day, it was cloudy.

**The Remarkable Observations of Tycho Brahe**

Further insight into the nature of planetary motions had to await better observations. The person who provided these observations was Tycho Brahe (1546–1601), a rather eccentric Dane. His eccentricities, however, served astronomy well. One of them was an obsession for eliminating all possible errors from his observations. He provided a long series of observations of planetary positions with the utmost accuracy of the time: one minute of arc (arc-min), that is, about one-thirtieth of the moon's diameter. We can appreciate his work better by considering that these data were gathered with instruments of sixteenth-century technology (or lack of it), and without benefit of a telescope. The data speak well of his perseverance and his fanaticism regarding the quality of his instruments. His instruments were in many instances solid gold and bankrolled by the Danish king (who also donated an island in fief for his support). The absence of a telescope demands instruments of very large size for the accurate measurement of angle. In Figure 2–5 we see a quadrant which occupied a whole west wall.

Tycho's work also included the observations of comets, and of a remarkable "new" star which suddenly appeared in the sky. Centuries later, present-day astronomers detected the remnants of an exploded star, a *supernova* (see color plate 16) very near Tycho's recorded position. Tycho also addressed himself to the question of whether the earth moved about the sun, and initiated careful measurements of the relative positions of several stars. He looked for a possible change in their configuration due to a change in perspective, as a result of the earth's annual motion. Unfortunately the stars seemed to remain stationary. As Chapter 7 will show, the apparent positional change of a star due to the earth's orbital motion never amounts to more than one second of arc (arc-sec), while Tycho's errors of measurement were at least sixty times larger. He concluded that the earth did not move, ignoring the remaining possibility: that the stars were much farther away than he could imagine. As a final note, history records that Brahe made one of his greatest contributions to astronomy inadvertently. This was his choice of a successor: Johannes Kepler.


## Laws of Planetary Motion


Johannes Kepler (1571–1630) was the inheritor of Tycho's enterprise, but fortunately for astronomy, he followed his own interest in *interpreting* the data already at his disposal, rather than devoting himself to observation. His principal effort was directed toward the planet Mars

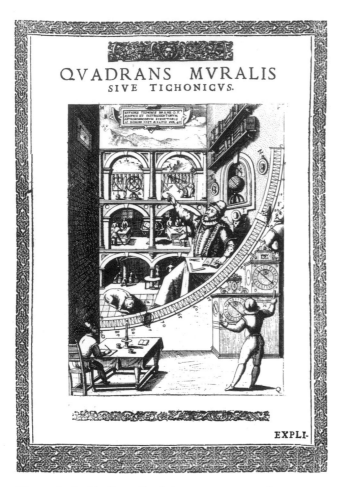

## QVADRANS MVRALIS
### SIVE TICHONICVS.

Figure 2–5. *Tycho Brahe's great mural quadrant. The large size promoted accuracy in the measurement of angles.*

for which Tycho himself had left an especially large amount of data. From this data, Kepler, who admired Copernicus, attempted to determine Mars' correct circular orbit about the sun. The mathematical task was enormous, even though some labor was saved by the use of the newly invented logarithms. Yet, after ten years of effort, all Kepler had to show for his work was failure. In his circular orbit, the calculated position of Mars could differ from its observed one by up to 8 arc-min. Yet, Tycho's observations had been eight times more precise, so Kepler could not bring himself to believe that the job was finished.

Eventually, his dissatisfaction led him to question the assumption of a circular shape for the orbit. Kepler cleverly found a method of obtaining the shape of Mars' orbit from Brahe's observations. Actually, what Kepler found was the distance of that planet from the sun relative to the earth's distance from the sun, at several points in the orbit, considering the earth-sun distance invariable. Fortunately, nature cooperated, for the earth has very nearly a circular orbit about the sun. In carrying out these measurements, Kepler was aware that Mars returns to any point in its orbit after a time interval slightly less than two years, actually 44 days less. Now, as shown in Figure 2–6(a), if the earth starts at position 1, it will be near

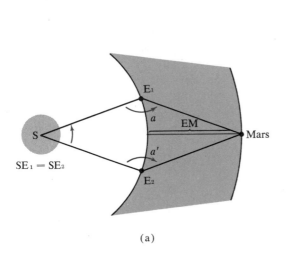

(a)

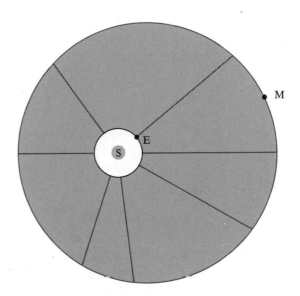

(b)

Figure 2–6. *(a) The distance from earth's assumed circular orbit to Mars' orbit is inferred from the shape of figure $SE_1$, $ME_2$. Angles* a *and* a′ *are directly measured;* b *is calculated. (See text.) (b) Applying the method of (a) many times yields the shape of Mars' orbit. Kepler immediately found it to be distinctly noncircular.*

position 2 after almost two years. The earthbound observer can measure angles *a* and *a′*. Angle *b* must be $44/365\frac{1}{4}$ of a circle, about 44°. Since any figure drawn with these angles must have the same proportions, the distance SM/SE is obtained. In Tycho's long years of observation, there happened to be many such observations from which Kepler derived the distance of Mars from the sun. From these, many distances SM could be found around Mars' orbit [Figure 2–6(b)]. Indeed, the orbit was not a circle, but its first cousin, an ellipse. Furthermore, the sun was not at the center of this ellipse, but at a *focus* (a point which plays a role analogous to the center of a circle, as shown below).

## About Ellipses

First, it is useful to notice the kinship of the circle, ellipse, and straight line. Observe a coin as it is tilted away from you. Starting from a circle, the rim of the coin takes on the shape of a continually exaggerated ellipse, and if the coin had no thickness, it would tend to a straight line as the tilt angle reached 90°. The term *eccentricity* is used to describe the amount of flattening of an ellipse. It is a number calculated from the long and short (major and minor) axes of the ellipse, and ranges from 0 for circles to exactly 1 for a straight line. The eccentricity for planetary orbits does not exceed 0.26. However, for the comets of the solar system (to be discussed later), the orbits are so considerably flattened that eccentricities very near 1 are the rule. The role of the focus can be seen in the construction of Figure 2–7. Bringing the foci together generates a circle, while separating them leads to the straight line.

## Continued Discoveries

Kepler had already found that the orbit of Mars was not circular in shape. Furthermore, he found the motion of Mars not uniform. At the closest point to the sun (perihelion) the orbital speed was greatest, while at the distant extreme (aphelion) it was slowest. Rather than yielding to befuddlement over these newly discovered irregularities, Kepler continued to scrutinize the nature of this motion. Finally, he found a new regularity, the *law of equal areas:* in equal intervals of time, the line joining the sun to the planet sweeps out equal areas. Figure 2–8 shows how this is related to the variable speed of the planet. In effect, Kepler was saying something *quantitative* about the speed variability.

Kepler next examined the data for the other planets, finding that they also traveled in ellipses with the sun at a focus, and obeyed the law of areas. From this common behavior for the planets, Kepler suspected that the period of a planet, the time needed to revolve once about the sun, was related in some simple way to its distance from the sun. Such a relation was found and is shown in Table 2–1. For each planet, the cube of the distance equals the square of the period, or $a^3 = P^2$ from direct observational evidence (see note on table). This is Kepler's harmonic (or third) law.

Kepler's great contribution, then, was the distillation of the apparently complex planetary behavior to three simple laws. This simplification process has remained one of the most fundamental aspects of science.

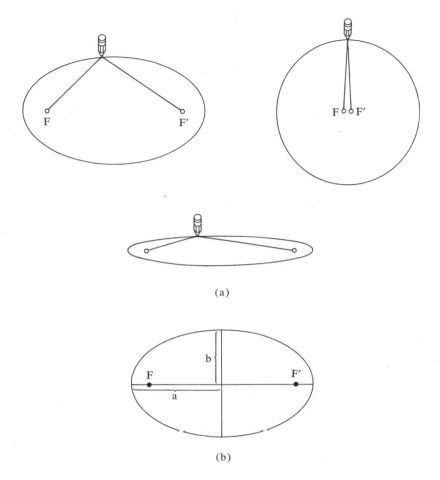

(a)

(b)

Figure 2–7. *(a) Drawing an ellipse with tacks and pencil. The points FF' are the foci of the ellipse. (b) The major axis (2a) and minor axis (2b) of an ellipse. For planets with elliptical orbits, the average sun-planet distance is* a.

## New View via the Telescope

Support for the heliocentric theory, among other revelations, was soon forthcoming from Galileo Galilei (1564–1642). These came through discoveries with a telescope, the first to be used in astronomical investigations. The first record of the telescope's invention appeared in Holland in 1609. Galileo fashioned his from its description, beginning observations

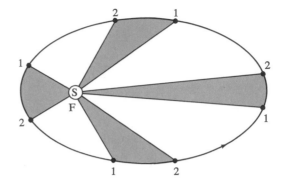

Figure 2–8. *Kepler's second law regarding motion in an elliptical orbit—the blue areas are equal if the time to go from 1 to 2 is constant. (See the text for details.)*

## TABLE 2–1

*The Observational Test of Kepler's Third Law*

| Planet | Distance from sun (a.u.) cubed | Orbital period (yrs.) squared |
|--------|-------------------------------|-------------------------------|
| Mercury | 0.0580 | 0.0580 |
| Venus | 0.3784 | 0.3784 |
| Earth | 1.0000 | 1.0000 |
| Mars | 3.5380 | 3.5380 |
| Jupiter | 140.8 | 140.7* |
| Saturn | 867.9 | 867.8* |
| Uranus | 7058.0 | 7058.0 |
| Neptune | 27160.0 | 27140.0* |
| Pluto | 61340.0 | 61340.0 |

\* The slight discrepancies in the values arise from the mutual attractions of the planets. The law is exact for two bodies.

in 1610. This "marvelous" instrument magnified about three times and could reveal stars ten times fainter than those visible to the unaided eye. Its optical quality would today be judged abysmal, but with it Galileo could see details on the moon's surface, the presence of spots on the sun's surface, four of Jupiter's moons, the moonlike phases experienced by Venus, and the resolution into stars of the luminous ribbon of the Milky Way.

   Far-reaching conclusions could be drawn from each of the observations. It could now be seen that the moon was a body much like the earth, with mountains and valleys, suggesting that planets too might be kin to the earth. The sunspot observations allowed the sun's 27-day rotation period to be discovered, and cast strong doubts on the popularly-believed unblemished perfection of the celestial bodies. The systems of Jupiter and its moons presented Galileo with Copernicus's idea in miniature. Furthermore, the observations of Venus' phases showed that one aspect of the Ptolemaic system was definitely wrong. Figure 2–9 indicates the varying appearance of Venus expected from both the Ptolemaic and the Copernican view. Galileo's observation that Venus' disc could be more than half illuminated clinched the Copernican version, because it meant that Venus traveled behind the sun, in contradiction to the Ptolemaic system. The Milky Way observations pointed to a universe considerably larger than had previously been considered. All these revelations made a sensational impact not only on Italy, but throughout Europe. Unfortunately, the new ideas butted abrasively with long-cherished notions. Galileo's view was held to be in contradiction to church teachings and resulted in his confinement to his home during the last decade of his life.

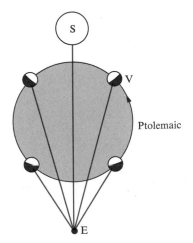

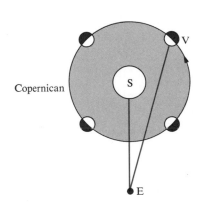

Figure 2–9. *The Ptolemaic theory says the view of Venus should always be a crescent. This is not observed. The Copernican view says Venus becomes near-full before being lost in sun's glare. This is observed.*

## Contribution to the Understanding of Motion

In addition to his observational interests, Galileo possessed a great curiosity regarding the nature of motion. His research demolished two widely-held assertions: heavy bodies fall faster than light ones, and the "natural state" of an object is to be at rest. The first was disproved by careful experimentation. The second required a simple inference drawn from experiments. Noting that objects sliding on ice traveled farther than on marble floors, given the same starting conditions, Galileo inferred that *in the absence of all friction,* an object could continue moving forever in a straight line. There was nothing natural about the state of rest, except that it could be naturally explained through the action of friction. Unfortunately, in an effort to comprehend the curved path of the planets, he mistakenly concluded that objects started in a curving path continue in a curving motion forever. This generalization is incorrect, since, as we will see below, curved motion requires the action of a *force.* His novel concept of *inertia* was fundamentally correct; however, it required that forces act anytime the speed of a body is changed. These investigations of Galileo set the stage for the complete understanding of motion soon to come.

### Understanding of Motion Completed

The capstone to the study of motion was put in place by a man often counted among the greatest geniuses who ever lived, Isaac Newton (1643–1727). Newton first adopted Galileo's concept of inertia, recognizing it as an innate property of matter. A body's intertia, he held, depended on the "amount of stuff making up the body," or its *mass.* The larger the mass, the greater the body's resistance to a change in its motion. Newton realized that all deviations from uniform, straight-line motion, due to changes of either speed or direction, require the intervention of a *force.* These alterations of motion, involving either a speeding up, slowing down, or change in direction are known as *accelerations* (Figure 2–10). The question was *how much* force would be required to produce a given acceleration on a given mass? Newton found the answer to be given by a simple formula:

$$\text{force} = \text{mass} \times \text{acceleration}$$

Figure 2–10. *Forces act to increase (a) or decrease (b) the speed of an object, or to change the direction of the motion (c).*

(a)

(b)

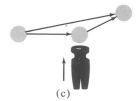
(c)

Newton further noted that forces always occur in pairs, oppositely directed (Figure 2–11). These ideas are known collectively as the three *laws of motion,* which are summarized below:

1. An object at rest remains at rest, or an object in motion remains in constant straight-line motion unless acted on by a force.

2. Force equals mass times acceleration.

3. For every force, there is an equal and opposite force.

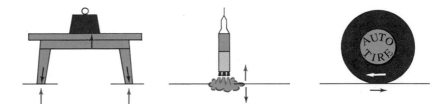

Figure 2–11. *Typical situations involving pairs of forces.*

## Comments on the Laws of Motion

The laws of motion involve the three quantities: *force, mass,* and *acceleration.* Let us say a few words of explanation about each.

Everyone has an intuitive feeling for a force as a push or pull, and this is, in fact, an adequate notion for our purposes. However, a force is not completely described unless information on the *amount* of force is supplemented by specifying the *direction* in which the force is acting; for example, a falling weight strikes the ground and exerts a *downward* force. A quantity such as a force is a *vector.* Another example of a vector quantity is wind velocity: the wind is blowing at 85 mph from the northwest. The amount and direction of the velocity are both given. It is likewise with accelerations. We might say, for instance, that a car accelerated 5 mph each second headed west.

On the other hand, quantities such as temperature, bank balance, or population have no notion of direction associated with them. They are called *scalars.* Mass is a scalar. We are only interested in how much of something there is. Mass is not to be confused with volume, which measures the size of something; nor is it the same as weight. The weight of something is a force, as we will see below. Remember that mass is a fundamental property of matter. The only way the mass of something can be altered is by adding to or deleting a portion of the something.

These remarks prompt a slight addition to the second law. A force acts with a given direction on a mass. *The direction in which the mass accelerates is the same direction as the force applied.*

## The Force of Gravitation

Suppose a mass is whirled at a constant speed at the end of a length of string. Since the *direction* of the motion is continually changing, a force must be continually acting and, indeed, is clearly perceived as the tugging force felt on the string. If the string is released, it may be observed that the object proceeds in the direction it was headed at the instant of release. We immediately learn from this humble experiment a lesson of astronomical import: *a force must exist which causes the planets to move in orbits about the sun.* Such consideration led Newton to propose the existence of a force of attraction or "gravitation" between every pair of bodies in the universe. The amount of the attractive force was to be given as the product of the two masses, divided by their separation distance squared:

$$\text{gravity force proportional to: } \frac{\text{mass 1} \times \text{mass 2}}{(\text{separation})^2}$$

The separation was to be reckoned from the centers of spherical bodies. It is interesting to note that when Newton proposed this formula, no laboratory experiment of the day was capable of verifying it, principally because laboratory masses produce forces so small compared to frictional forces that they are difficult to measure without modern techniques. However, the laboratory used by Newton to verify his proposed formula was nature itself—the solar system bodies move without friction and their motions are dependent only on gravitational forces. We may note that this need to make use of "natural experiments" has remained one of astronomy's distinguishing features as a science, and is a direct result of astronomy's dealing with nature on such a large scale. In this case, following the example of Newton, we will verify the correctness of the numerator and then of the denominator of the gravity formula from observations of nature.

Before Newton, Galileo had carried out experiments on falling bodies. He found that falling objects of *differing* mass behaved *identically:* they all fall with the same acceleration, increasing their speed by 32.2 ft/sec each second (980 cm/sec each second) in the absence of friction. However, according to the second law of motion the acceleration must be given by $a = f/m.$ In this case, the objects fall because a force is supplied by the gravitational attraction of the earth's mass for that of the falling body of mass $m$. If we substitute the gravity formula for $f$, we have

$$a = \frac{f}{m}, \text{ proportional to } \frac{m_{earth} \times m}{(\text{distance to earth center})^2} \times \frac{1}{m}$$

In order for the acceleration to actually be a constant, the mass of the object would have to cancel, and this could occur only if the gravitational formula contained the masses of both the earth and the object as *factors*. Looking again at the gravity law, we can verify that the acceleration of any object will be lessened if it is moved to a greater distance. It occurred to Newton that the moon was an ideal object to test the distance dependence of the acceleration of a falling body. Located some sixty times farther away from the earth's center than objects at its surface, the moon's acceleration toward the earth should be only 1/3600 (1/60 × 1/60) of 32.2 ft/sec per second, or 0.0089 ft/sec per second (.21 cm/sec per second). Newton was able to show, on the other hand, that the *actual* acceleration of the moon toward the earth, caused by its constantly changing direction in its orbit, could be calculated simply from the knowledge of the moon's distance and orbital speed alone. The calculated value was 0.0089 ft/sec per second, in precise agreement with a force diminishing as the square of the distance from earth's center.

The discovery of the gravity formula marked one of the great early theoretical advances of the physical sciences. It has been successfully tested countless times since its discovery, earning the appellation "law of gravity."

## Consequences of the Gravity Law

### Masses of Astronomical Objects

Newton himself began the search for the consequences of the gravity law by investigating the motion of two mutually attracted objects. He found that Kepler's three laws could be obtained mathematically from his gravity law, and that the correct form of Kepler's third law was:

$$\frac{a^3}{P^2} = \text{sum of the masses of the two bodies}$$

The correct units to use in this equation are: astronomical units for $a$ and years for $P$, a which yields the answer in *solar masses* (sun = 1 solar mass). The sun constitutes over 99 percent of the mass of the solar system, so if we apply the formula for the case of a planet orbiting the sun, then the sum of the masses will be very close to 1, which agrees with Kepler's original third law. The new form of Kepler's third law has a very useful property: the left side and right side of the equation are the same numerically, of course, but the left side is *observable,* while the right side is not. In other words, we can find the *total mass* of any two orbiting bodies by inventing techniques to measure *a*. (Only a clock is needed to find *P*.) For example, we can measure in this fashion the mass of a planet which has a satellite. The distances to the planets are known, so if we measure the angle between one and its satellite, we can then obtain the separation distance, using astronomical units for this length (Figure 2–12). Likewise, we can easily clock the satellite's period of revolution. Let us express this in years. Thus we find $a^3/P^2$, the sum of the masses, in units of the sun's mass. This total mass is composed almost entirely of the planetary mass. The moon's mass was actually measured this way in the 1960s by placing artificial satellites in orbit around it. The method has even been refined to the point where the mere deflection of an interplanetary rocket by a solar system body can be used to find that body's mass.

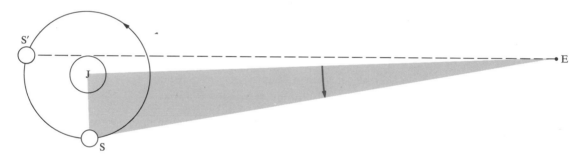

Figure 2–12. *With Earth-Jupiter distance known, observation of the angle* a *yields the distance JS by trigonometry. We determine the satellite's period by timing its disappearances behind the planet. Taking care to use consistent units, we can determine the sum of the masses of the bodies. In this case, we effectively determine Jupiter's mass.*

There is no need to limit our horizons in determining masses to objects in the solar system, because any pair of orbiting bodies is susceptible to this law. A perusal of the sky with binoculars shows, for instance, many close pairs of *stars* that might be candidates. Many years ago, the astronomer William Herschel proved that too many of such pairs existed for all of them to be chance coincidences—at least some had to be bound by gravitation and obey the third law. Thus, we may obtain information on stellar masses, too. The results are of great importance and will be discussed in more detail in Chapter 8.

## Orbital Shapes

Newton also painted a more complete picture of the orbital shapes assumed by two gravitating bodies. Let us first consider the case where one of the two objects is very much less massive than the companion, for example, the sun and a planet. A planet could travel in a circular orbit, Newton showed, only if it had an appropriate orbital speed which depended on its distance from the sun. The farther the planet was from the sun, the less was this required speed. If an orbiting body's motion were different from this "circular speed," its orbit would be an ellipse. Figure 2–13 shows the situation: bodies with speeds smaller than the circular velocity "fall in" toward the sun during their orbit, returning to their starting point, which becomes the most distant point of the orbit (*aphelion* point); bodies with speeds exceeding the circular velocity will "fall away" from the sun during their orbit, returning to their starting point, which becomes the nearest point of their orbit (*perihelion* point). Not all orbits close, however. For example, if the speed of a body reaches $\sqrt{2}$ times the circular velocity (about 141 percent greater), the orbit fails to close. The body leaves on a path known as a *parabola*. The object is said to have reached the "escape speed." Larger speeds likewise produce escape; however, the escape is faster, and the body leaves on more open paths known as *hyperbolas*.

A number of interesting observations may be made concerning orbital shapes found in the solar system. First, all the planets of the solar system revolve in the same direction with nearly circular orbits. This is an important clue to their origin, as we will see in Chapter 5. On the other hand, the comets of the solar system travel in very flattened ellipses which reach out typically to the far regions of the system.

Ellipses of modest shape are utilized in space travel to move from the vicinity of one planet's orbit to that of another (Figure 2–14). To travel to Mars from Earth, for example, a spacecraft is injected into an orbit with a speed slightly greater than the circular speed, after first having gotten clear of the immediate gravitational influence of Earth. The orbit has the same sense of revolution as the earth's orbit, receiving, in effect, a free boost from the earth's circular speed. On the other hand, to go to Venus, we need to move about the sun slower than the earth does to swing into an orbit bringing us closer to Venus. Interplanetary travel is also subject to one further consideration: the craft's arrival at the planet's orbit should coincide with the arrival of the planet there as well!

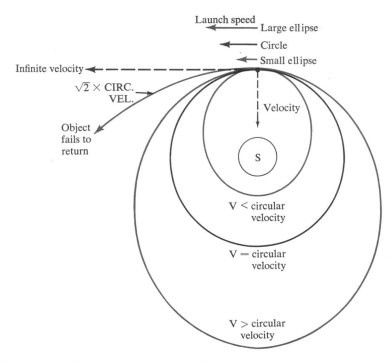

Figure 2–13. *The shape of the orbit depends on the speed of injection into orbit. The speed for a* circular *orbit decreases for bodies farther from the sun, so that each distance has its specific circular speed. A speed greater or less than the circular velocity appropriate to a given distance will yield an orbit of a different shape.*

### Two Bodies of Comparable Mass

The case of two orbiting bodies of similar mass occurs widely in nature as pairs of gravitating stars. We are fortunate to be able to observe some of these as "visual binary stars" in which we can actually record photographically the orbital motion, seen against the background of the stars. The path of the stars is in general elliptical, and both stars move. To make the reason for this plausible, remember that the gravity force is mutual. Both objects are subject to the same amount of force, but the one with the larger mass has the greater inertia and thus will have the smaller net motion. As a practical analogy, consider the case of a large and a small person on a seesaw. The one with the greater mass sits closer to the

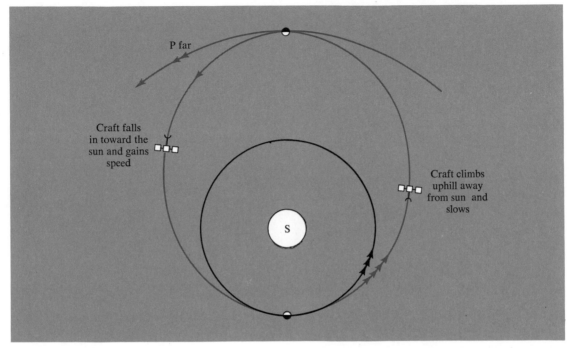

Figure 2–14. *Interplanetary travel with the least expenditure of energy uses ellipses just touching both orbits. It is also the path least economical of time. Faster, more direct travel requires a great deal more energy. (The number of arrows is a rough indication of relative speed.)*

balance point. As shown in Figure 2–15, the orbital motion of two stars has this same balancing property; the orbiting stars always lie on a line crossing the balance point, with the more massive star lying closer to it. As shown in Chapter 8, this balancing act can allow the *individual* masses of the members of a binary star system to be found.

Stars are so distant that only those pairs separated by many astronomical units can be seen as two separated points of light in the telescope; that is, as *visual binary stars*. (Closer pairs merge into apparently single images for reasons to be discussed in Chapter 5.) Kepler's third law implies that visual binary stars, because of their generally great separations, must take tens or even hundreds of years to complete their orbits. Figure 2–16 shows the orbit of a very short period visual binary and a portion of a very long period orbit. A famous visual binary, which can be seen by persons with acute vision (and easily with the aid of binoculars),

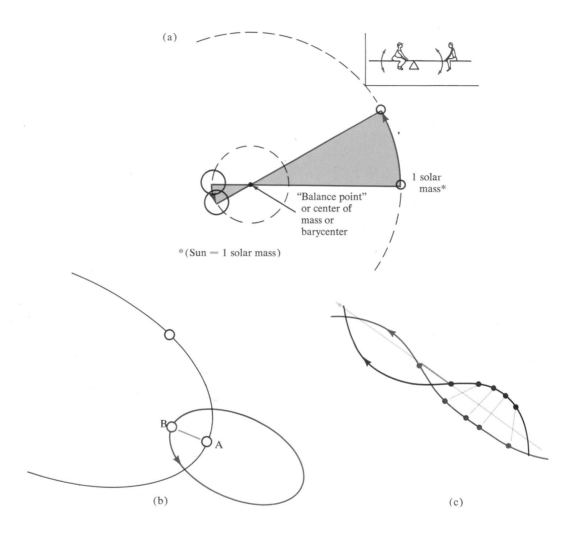

Figure 2–15. *(a) The circular orbit of two stars in a binary star system. They move orbitally as if the line connecting them were rigid and pinned to the* barycenter *(center of mass). Nothing constrains the binary system to be fixed to a spot in space; generally, it will move uniformly and slowly with respect to the other, distant stars, with the barycenter describing a straight line. The barycenter may also be inside one of the bodies; for example, that of the earth-moon system lies within the earth. (b) The elliptical orbits of two stars in a binary system. The mass of A is two times the mass of B. Ellipses have same shape and both components lie opposite the barycenter at all times. (c) Path on the sky of two stars in binary system with circular orbits, seen against the background of stars. The barycenter moves on a straight line.*

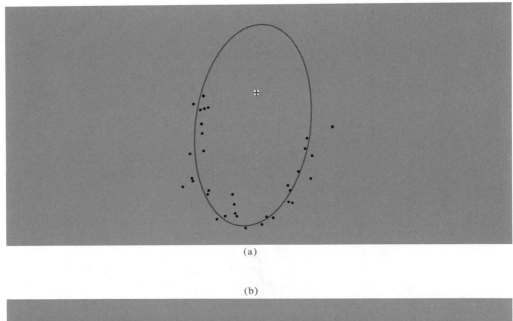

(a)

(b)

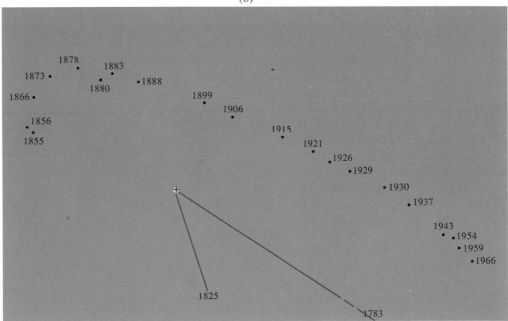

Figure 2–16. *Visual observations of a binary show the effects of their mutual gravitational force. Visual measurements with respect to the brighter star of the pair (white dot) are shown here for the short-period binary 13 Ceti (a), and the long-period system 40 Eridani (b). The short period binary is a very difficult object to observe, due to the very close proximity of the stars.*

is Mizar and Alcor—the double star at the bend of the handle of the Big Dipper. Furthermore, the brighter component of the visual double is itself a very close binary, too close to be seen as double in the largest telescope. Such very close pairs may still be discovered to be *spectroscopic* binaries, as will be discussed in Chapter 8.

### Three Is a Crowd

The problem of the motion of two gravitating bodies was quickly solved by Newton. On the other hand, the case of three objects became one of the most vexing and celebrated problems of dynamical astronomy. So far as is known even today, it is impossible to find a general *formula* which predicts the motion of three bodies when given their beginning situation (their initial positions, speeds, and directions of motion). A few special cases are exceptions to this rule and are interesting enough to warrant mention. Many years ago, a gentleman famous for his mathematical legacy, J. L. Lagrange, found that under certain conditions three bodies at the corners of an equilateral triangle (three equal sides and equal angles) would remain so situated. For example, construct such a triangle, placing the sun and Jupiter at two of its corners. Figure 2–17 shows that the third corner lies on Jupiter's orbit, preceding and trailing the planet. This suggested that any particle reaching these points with speed near Jupiter's orbital

Figure 2–17. *Relatively lighter objects can be stably located on the vertex of an equilateral triangle, with the two massive bodies at the other corners. For the Jupiter-Sun system, small asteroidal bodies are actually found at the leading and trailing points, as shown.*

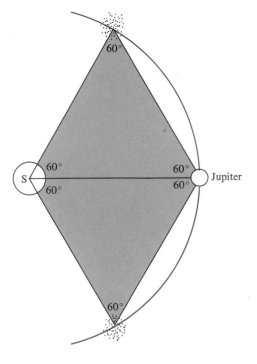

speed might be likely to become trapped and stay there. Astronomers therefore searched the sky near both of these "triangular points" and were rewarded by the discovery of small bodies which had become trapped there. About a score of these small bodies are known now as the Trojan group of asteroids, or minor planets. Since every system of two bodies should have such points, visual observers have recently attempted sightings at the earth-moon triangular points. The discovery of very faint luminous clouds has been reported, but confirmation is lacking since they are extremely difficult photographic subjects.

It is interesting to note that nature exhibits a pronounced bias in arranging three-body stellar systems. Invariably, a visual triple star system consists of a close pair and a third distant companion.

**Answers Without Formulas**

Multiple systems of stars ranging from four, five, and more members up to whole clusters of stars are also known. (See color plate 10.) Furthermore, galaxies consist of billions of stars. No formulas are available for the motions of these bodies; it would be hopeless to even search for them.

At this point we should distinguish between the ability to write *formulas* for the motions of several gravitating bodies and the ability to follow their motions by *direct calculation*. Fast computers have now made the technique of direct calculation practical for up to a few hundred thousand bodies, representing stars. The basic idea behind the computation is simple: for each body the gravitational forces of the rest of the group on it are evaluated, then allowed to accelerate that body. This results in new positions for each body. The process of finding the forces on each body can then be repeated. As the process proceeds, the dynamical evolution of the bodies unfolds. Such a brute-force technique for following the motion of the bodies is known as the method of *special perturbations*. One expert has even made computer-assisted animated films illustrating the results of his investigation of simulated galaxies. Another example of the usefulness of this computer method is an inquiry into the very long-term future of the solar system, posing the question of whether the planets will remain orbiting the sun in the same fashion as at present over the next five billion years or so. The answers so far indicate that the solar system will change little over that time.

## Tidal Effects

If one spends some time relaxing at a coastline, the machine-like regularity of the ocean tides is bound to be noticed. The sun and the moon provide the motive power for this great machine. However, they can act only through gravitational effects. How does gravity cause the rising and falling of the surface of the ocean? This question will be answered as soon as we introduce the fundamental idea of a *tidal force,* and examine its consequences.

### Tidal Force

Associated with every mass is a gravitational force which decreases with distance. In Figure 2–18, we have placed two masses connected by a rubber band at some distance from a large mass. The closer mass experiences the greater force of attraction, so if the distant mass is anchored, the rubber band is stretched. The difference in the gravitational forces on the two masses is called a *tidal force.* Thus, any object placed in the gravitational influence of another will experience some stretching, and the exact amount depends on the material making up the mass—steel stretching less than rubber, for example. The stretching tendency increases very quickly as an object is moved closer to the large mass; for example, moving three times closer yields a twenty-seven times ($3 \times 3 \times 3$) greater stretching force. In general, moving $n$ times closer results in an $n \times n \times n = n^3$ times greater tidal force. Is it possible that this rapid increase in tidal forces could result in the destruction of an object by a large mass? For example, could a satellite be disrupted by a perilously close passage to its parent body? This question is made more than academic by the existence of Saturn's

Figure 2–18. *Tidal effects resulting from a gravitational force which decreases with distance. The decrease in gravity farther from the massive object is responsible for a stretching effect. A "hand" restrains the dumb-bell test objects joined by a rubber band. What restrains the planets P and P'?*

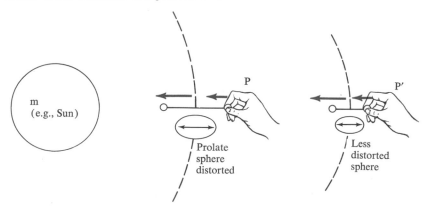

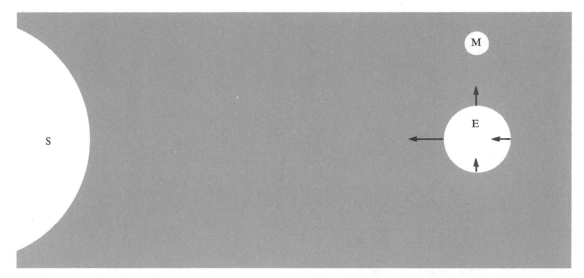

Figure 2–19. *Gravitational effects of the sun and the moon on the earth. The moon's force is smaller, but it* changes *more over the diameter of the earth than does the gravitational force due to the sun. The moon's tide-raising effect is just about double that of the sun's.*

rings, for they are composed of myriad small particles. In the 1800s, E. A. Roche showed that *tidal break-up* could indeed occur, if a satellite came within 2.44 planetary radii. Then the size of the tidal forces is greater than the size of the gravitational forces holding the satellite together. The rings of Saturn lie within 2.44 radii.

### Earth Tides

The *body* of the earth has a barely detectable tidal bulge (about twelve inches). It is due primarily to the moon, but in fact the sun also produces tidal effects with about half of the moon's efficiency. Even though it is considerably more distant than the moon, it is much more massive (Figure 2–19). The tidal effects of these two bodies combine as shown in Figure 2–20.

### The Ocean Tides

In contrast to the very small deformation of the earth's body, the gyrations of its liquid covering (the oceans) are quite obvious. The tidal effects on

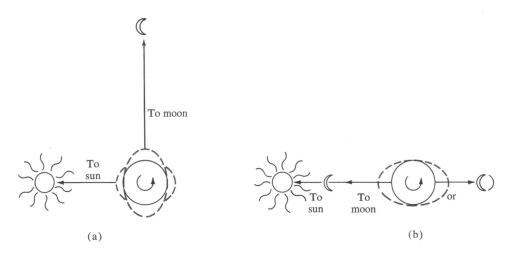

Figure 2–20. *Tidal effects of the sun and moon simultaneously stretch the earth, and cause a flow of the open waters of the ocean as shown above (greatly exaggerated). We can visualize the rotating earth as a kind of tank-tread. When the sun and moon are aligned, as in (b), the effects roughly add. Note that, at least qualitatively speaking, the tides depend on the phase of the moon. Ocean tides in (a) are "neap tides," while those in (b) are "spring tides." (See text.)*

the water, known as "the tides," are of a type different from that on the body of the earth, simply because water is free to flow while the land is not. Let us examine in detail the influence of the moon. Figure 2–21 shows the way forces are felt at the earth due to the moon. Since the earth and moon are not approaching or receding, but have reached a gravitational balance, let us remove the average force exerted by the moon on the earth's center from the forces illustrated in Figure 2–21 and see what is left. What is left is in fact the tidal force. We can see why the earth stretches along the moon's direction in response to these forces. Forces exerted along the earth's surface can and do move the water located there. We can see that *two* bulges tend to form from this pile-up, lying on a line pointing in the moon's direction. As the earth revolves each day, we are carried through the bulge twice, experiencing two high and two low tides due to the moon.

As we have noted, however, the sun is about half as effective as the moon in raising tides. In order to complete the picture of ocean tides, we can add the effects of both bodies, as in Figure 2–20. We see that the tides

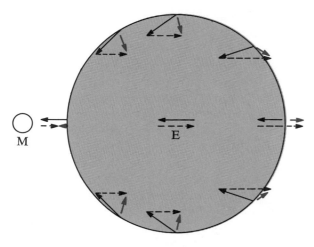

Figure 2–21. *Tidal forces (blue arrows) are the forces left over when the average force (dashed arrows) is subtracted from the total force (black arrows).*

are maximum when the sun and moon are aligned, as at new and full moon, when their tide-raising influences reinforce. Minimum tidal ranges are experienced when the moon is at a right angle to the sun-earth line, at the quarter phases.

These simple ideas are useful in understanding much of the grosser properties of the ocean tides. However, not all features of the real tides satisfy this picture. For instance, high tide at a certain location should occur when the moon passes overhead. This is seldom the experience at coastlines, where the high tide may be delayed for hours. Usually, this is understood by realizing that shallow areas tend to impede the free flow of the liquid. However, at certain spots, even in the open ocean, only two tides a day are experienced rather than four. Nobel prizewinner Dr. Hannes Alfven has suggested that our picture of the tides is not complete without considering that the tidal forces cause the waters of the oceans to have a sloshing motion, like the water in a bathtub. This sloshing can help account for the observation of two daily tides.

The fluid tides dissipate about two billion horsepower continuously as heat due to friction. After many millions of years, this exerts profound influences on the earth-moon system. A somewhat longer month and much longer day (equal to the lengthened month) are in store for us. None of us, however, are destined to see the end of moonrise and moonset; it is billions of years in the future.

## Condensed Matter

Later we shall discuss the nature of a star. One of the great astronomers of the past, A. S. Eddington, noted that a star is but a mass of heated gas which can resist the incessant order of gravity to collapse only so long as it can produce energy in its center. But no fire burns forever; collapse has to come ultimately to every star. The shrinkage is halted, however, when the material acquires enough stiffness to counter it. So far, astronomers know of three possible outcomes of gravity's working on dying stars. Stars like the sun stop shrinking when they reach the size of the earth because a new type of pressure, in addition to ordinary gas pressure, comes into play. These dying stars have been observed; they are the *white dwarfs.* The density of matter in a white dwarf is nearly two tons per cubic inch. The theory of stellar structure further informs us that stars more massive than the sun may encounter a situation in which they are not supported by this pressure. Their inner core region may shrink to about fifteen miles in diameter, reaching such densities that the *atoms* touch each other! Their density would be about a hundred trillion times that of water, and the surface gravity of such a body would be immense. The energy which could lift an object on earth to the height of Mt. Everest would suffice to lift it only one-third of one-millionth of a foot on such a body. These stars are called the *neutron stars,* which are very probably identified as radio-emitting objects called *pulsars.*

One last incredible scenario has been provided by the theoreticians. It appears possible that the stiffness afforded a collapsing body by the atoms touching may not, in all cases, be enough to stop the collapse. What then? No further state of matter is known which can provide a defense against further collapse. Collapse must go on forever! Ultimately, the gravity at the surface of the body becomes so large that even light is prevented from leaving it. This is the *black hole.* No one yet knows with certitude if such objects actually exist, but there is sufficient evidence now from studies of very peculiar close pairs of stars to be very suspicious. In Chapter 9, we will see how stars reach these remarkable terminal stages of life.

## Questions

1.  What is the principal difference between the Ptolemaic system and the Copernican system?

2.  What *observation* first weakened the case for the Ptolemaic theory?

3.  What are Kepler's three laws? What do they have to do with finding masses of celestial bodies?

4.  How did Galileo's explanation of the circular motion of a planet differ from Newton's?

5. What is meant by a tidal force? What bodies are most important in raising tides on the earth? How do ocean tides differ from land tides?

6. We have ignored the effect of tides on the moon itself. Could we compare them to the tides suffered by the earth? (Hint: The sun will affect the moon equally as it affects the earth—what other body would exert an important tidal influence on the moon?)

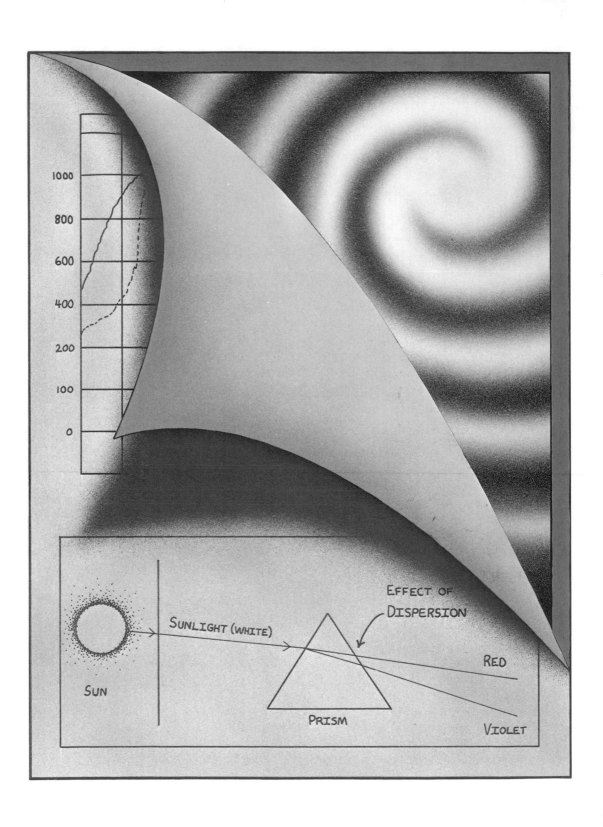

1000
800
600
400
200
100
0

EFFECT OF
DISPERSION

SUNLIGHT (WHITE)

RED

SUN

PRISM

VIOLET

# 3 Matter, Light, and Stars

## Introduction

In the previous chapter, we became acquainted with some of the successes of dynamical astronomy, which put into an ordered framework of simple laws the apparent arbitrary motions of the celestial bodies. However, an understanding of the stuff of which the stars are made, the revelation of their changes with time, or a knowledge of the extent, nature, and evolution of the universe requires a different kind of inquiry. To know what stars are, we have to ask, What is matter? and What is energy? We will see that even a simple understanding can be rewarding, giving us a real insight into the nature of the stars, and will carry us a long way toward an appreciation of the universe.

Matter and energy are the two fundamental constituents of the universe. Matter exists in three forms known as *states:* gaseous, liquid, and solid. On earth, the prevalence of liquid and solid states gives a false sense of their importance. Actually, in the universe at large, the gaseous state constitutes the overwhelming bulk of matter. This matter is mainly in the form of stars; they are gaseous. This point emphasizes that we should not accept our everyday impressions regarding matter without second thoughts. It also explains our future references to gases and matter generally, as if they were interchangeable.

Energy, too, has different forms. It appears as light or heat, or as sound, or an energy of motion, whether it is the motion of an atom, a molecule, or a machine. Energy may also be stored; for example, the electrochemical energy of a battery, or the chemical energy of fuel in a tank. It can also be stored in mechanical fashion; for instance, a weight lifted above the ground. When the weight falls, it can be used to run a clock or drive a pile. The many different forms of energy are also convertible into one another. Some examples are a battery operating a lamp, changing electrochemical energy into light and heat, or a falling weight hitting the ground, deforming and heating it, and creating sound waves. When energy travels, it often does so in the form of a wave. The focus of the following discussion will be on a traveling form of energy—radiant energy (for example, light)—because what is known about the stars comes from analysis of their radiant energy. All radiant energy originates with matter, and consequently analysis of the radiation can reveal much regarding the type and conditions of the radiating matter.

For the astronomer, light is literally a messenger from the stars. The message is in a kind of code called a spectrum (see Figure 3–1). An understanding of the nature of light is needed if we are going to make sense of

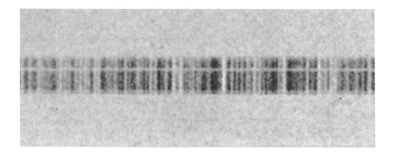

Figure 3–1. *A message from the stars.*

this message. However, the reward for deciphering these natural hieroglyphics is great indeed. Without journeying to the stars, we can learn of their nature, determine their chemical compositions, their masses, sizes, distances, their true brightness and motions, as well as their origins and fates. However, we will also find that light places limits on our investigations: it prevents us from seeing the details on the surface of any star but the sun. In even the largest telescope, the image of a star can look only like the images shown in Figure 3–2.

## The Nature of Light

### Wavelike Properties

Light is a form of energy, an example of the type known as radiant energy. All radiant energy travels at the same speed as light, which in a vacuum is 299,793 km/sec (186,300 mi/sec). Its speed is somewhat slower in denser media such as glass or water. Other examples of radiant

Figure 3–2. *Telescopic photograph of the image of a bright star.*

energy are X-rays, ultraviolet rays, infrared rays, radar, television, and radio waves. All of these examples differ among each other in the same way that, for instance, red light differs from blue light. A simple explanation for these differences will emerge when we consider that light shows many of the characteristics natural to *waves*.

Light's wavelike character is very similar to that of water waves; thus an illustration of the properties of water waves is useful. Waves can be created by oscillating a flat slab vertically on the water surface, causing waves to ripple away from it. The launched waves travel out from the source, disturbing the surface of the water. The *vertical* bobbing of a floating object, such as a cork, assures us that it is only the *disturbance,* and not the water surface layers themselves, that is propagated. The rate with which the slab is oscillated governs the length of each wave, or its *wavelength.* Rapid oscillations of the slab give rise to waves of short wavelength, compared to those arising from slow oscillations, as seen in Figure 3–3. The number of oscillations per second is known as the *frequency.* For each complete oscillation of the slab, one wave is created. At the next oscillation a second wave is formed which smoothly continues the first. Now suppose that in one second, $f$ oscillations occur (the frequency is $f$) and each produces waves of length $l$. Then the first wave has traveled $f \times l$ distance in one second. This amount of distance covered in a second is the speed of the wave. After all, speed is just distance covered per unit of time. Since this observation regarding wave propagation would follow for waves no matter what the type, we can write for all types of waves:

$$\text{frequency} \times \text{wavelength} = \text{speed}$$

It is important to realize that wavelike motions carry energy. This is evident just from watching the lifting and pitching motions of boats in a harbor. It surely requires energy to lift these masses. Furthermore, it is possible to recover some of this wave energy and make use of it, as shown in Figure 3–4.

## Diffraction

A peculiarity of waves which will interest us even more later is their behavior at obstacles. In Figure 3–5, the situation is illustrated by the

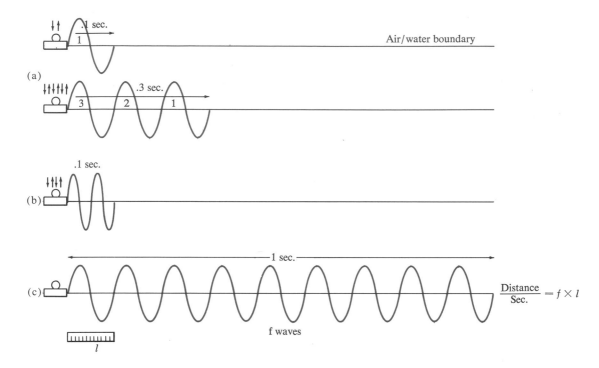

Figure 3–3. *(a) Waves produced on an air/ water surface propagating away from the source. (b) A faster rate of oscillation of slab produces* shorter *waves. The frequency of the wave in (b) is twice that of the wave in (a). (c) The velocity of a wave is always the number of them produced each second times the length of one wave.*

behavior of a sea wave impinging directly on a sea wall. Although it might be suspected that there exists a shadow region behind the sea wall entirely free of waves, waves actually are found to spread at the edge of the sea wall, and propagate into the region of the geometrical shadow. This spreading at obstacles is called *diffraction*.

   We can set up an analogous optical experiment to test for wavelike behavior of light. Here, the sea wall becomes an opaque screen and photographic film is placed behind it to see if any light waves diffract around the screen. From Figure 3–5(b) there is no doubt that they do. Since diffraction always occurs for waves (sound waves, too), we may be assured of light's wavelike properties.

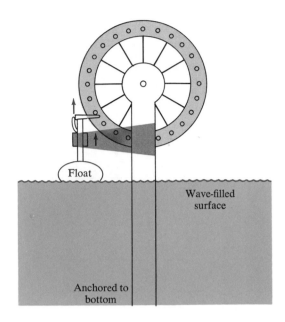

Figure 3–4. *Preliminary sketch of machine for extracting energy from waves.*

## The Doppler Effect

Another intriguing wave effect is the apparent frequency change which occurs when the wave source moves with respect to the observer, or vice-versa. For example, in Figure 3–6, making use of water waves once again, we can imagine we are at rest as the waves approach, and we count the number of waves which travel by us each second, thus determining their frequency. If we now approach the source, traveling *against* the direction of the waves, this frequency *increases*. If we move *away* from the source, traveling with the direction of the waves, the frequency decreases. This phenomenon, known as the Doppler effect, is common to all wave motions. Most people notice it in sound waves—the apparent sudden decrease of pitch noticed in a bell or whistle, when moving away from the source. A frequency change also occurs in light and all other waves under relative motion of the source and observer. A case of particular interest to us is the Doppler effect on the light emitted by stars, moving with respect to the earth.

## Light and Related Waves: The Spectrum

Light is a disturbance with wavelike properties. Analogous to water waves, something must be shaken or oscillated to generate such a wave. Elec-

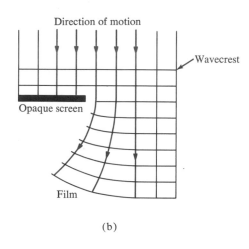

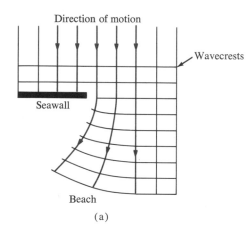

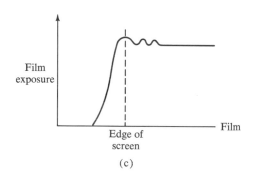

Figure 3–5. *(a) The bending of water waves around an obstruction. Waves are "diffracted" into the shadow zone. (b) The similar diffraction of light waves about an obstacle. (c) Photographic effect of bending of light.*

trically charged particles known as *electrons,* which we will see are a fundamental constituent of matter, give rise to electromagnetic waves when oscillated. These waves, however, require no medium for their propagation. They are naturally "self-propagating," and have no difficulty

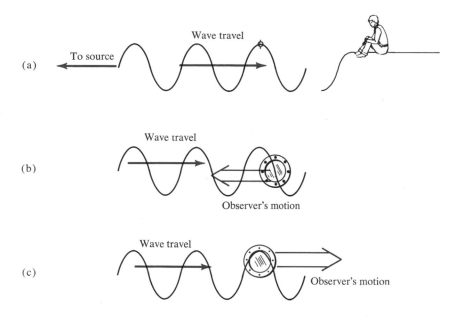

Figure 3–6. *(a) Wave watcher at rest counts crests per second by observing bob. (b) Same wave, as seen through the porthole of a ship moving toward a source. This observer would count a higher frequency of waves per second. (c) Observer traveling away from source notes lessened frequency.*

traversing even a perfect vacuum. The wavelengths of visible light range from $7 \times 10^{-5}$ cm for deep red light to $4 \times 10^{-5}$ cm for deep blue light. Between these extremes lie all the colors of the spectrum: *red, orange, yellow, green, blue, indigo,* and *violet* (mnemonic ROY G. BIV) (Figure 3–7). X-rays, ultraviolet waves, heat or infrared waves, radio waves, and all other forms of radiant energy differ from light only in the wavelength of the radiation. Radiant energy is conveniently ordered by wavelength in a *spectrum,* and the complete spectrum is shown in Figure 3–8. Virtually all the regions of this spectrum are now studied in astronomy, as a result of recent technological advances. After many centuries of effort confined to the visible spectrum, work is currently under way to detect celestial sources in many other wavelengths. So far, the greatest attention has been given to the radio region, but recently infrared and X-ray observations have been vigorously pursued, yielding a wealth of information and surprises. Wavelengths longer than infrared are used in radio communication, including AM, FM, television and radar. The techniques for working in different regions of the spectrum differ considerably. Wavelengths in the

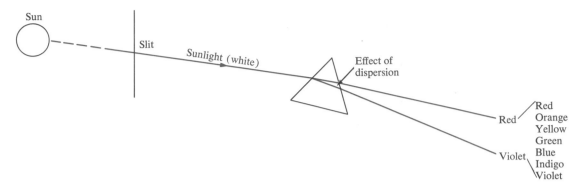

Figure 3–7. *The spectrum of sunlight. Newton showed that a simple prism could be used to disperse light into its component colors. In other words, a prism sorts the radiation by wavelength, forming the basis for its quantitative examination. The analysis of a spectrum is a technique common to all of the physical sciences. The dispersing element is the basic component of a spectrograph, to be discussed in Chapter 5.*

spectrum accessible to FM radio in the range of 2.7 m to 3.4 m, for example, are selected electronically. If we have a prism, however, we may select wavelengths in the visible spectrum merely by sliding a slot across the spectrum.

Let us note that it is an old practice to specify light wavelengths by taking the wavelength in centimeters and moving the decimal point *eight*

Figure 3–8. *The electromagnetic spectrum.*

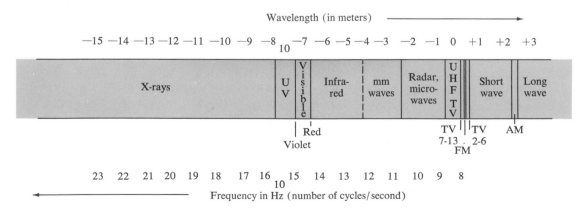

places to the *right*. Thus, yellow light with wavelength $5.5 \times 10^{-5}$ cm, or 0.000055, becomes 5,500 Å. The Å stands for Ångstrom, the name of a distinguished Swedish scientist, and the number is read "5,500 angstroms." In the following discussions we will use these units for light wavelengths. Continuing the example, the boundaries of the visible spectrum are from 4,000 Å (deep blue) to 7,000 Å (deep red). As an exercise, try calculating the frequency of visible light using the general formula above; be prepared for a large number.

## Particlelike Properties

In this section we will see evidence that light's behavior is too complex to be totally explained by a wave picture. Instead, there is evidence that light simultaneously has properties which we normally ascribe to particles. This may seem paradoxical at first, but we should remember that our common ideas about matter are based on the behavior of objects on a much larger scale than a size of light wavelengths, so familiar concepts might prove lacking at such a small scale. Such is evidently the case.

While the evidence of diffraction assures us that light behaves as a wave, it fails to explain the following experiment. Let us perform a very simple experiment in which a beam of light is allowed to fall upon a light-sensitive electronic device, a phototube which converts light into an electric current. The tube is connected to an amplifier which in turn is connected to a meter, as shown in Figure 3–9. With a bright beam illuminating the tube, we find the meter giving a quite steady reading. Upon greatly reducing the intensity of the light, and increasing the sensitivity of the device, we will come to a point when the meter's needle will move irregularly about its average reading. Continuing in this way, we will find, remarkably, that at extremely low light levels (well below that which is

Figure 3–9. *An experiment on a light beam is performed with ideal devices. See text for explanation.*

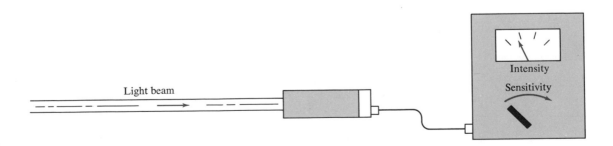

perceptible by the eye) the needle will rest at zero, but will also execute spasmodic jumps at irregular intervals. The significance of this is obvious —light is not continuously divisible. Evidently, our light beam consists of numerous particles, each traveling along at the speed of light and each endowed with a bit of the energy the beam transports. These particles have been given the name *photons*. Each photon carries a specific amount of energy, and this amount of energy is found to be proportional to the frequency of the light (or, proportional to 1/wavelength). This means that if we have two streams of photons, one of red light and one of blue light, the photons in the blue beam *each* carry more energy. (Of course, the total energy in each beam depends on the number of photons in the beam.) Likewise, we would find that photons of ultraviolet light carry more energy than photons of blue light. For this reason, the burning effect of sunlight is due to its ultraviolet rays. The ultraviolet light photons each have so much energy that they can initiate irritating chemical reactions in the skin. When the wavelength of the radiation is short enough, as in X-rays, the photons have a great deal of energy, and can penetrate the body and expose photographic film.

## The Nature of Matter

The real construction of matter is hidden from casual view because its elementary building blocks are very small. Thus what the eye or the sense of touch perceives as a smooth and continuous piece of material (for instance, aluminum) is found upon closer examination of direct and indirect evidence to consist of submicroscopic objects known as atoms (atoms of aluminum). Aluminum is an element, one of ninety-two occurring naturally. Every element is composed, if pure, only of atoms of one type. The simplest element, in terms of its atomic structure, and the one most abundant in the universe is hydrogen. It is a flammable gas at ordinary temperatures and pressures. The second most simple element in complexity is helium, which is also the second most abundant element in the universe. It is a nonflammable inert gas. Lithium is the third most simple element, a pasty solid. However, lithium is not third in abundance, nor are the remaining elements' abundances necessarily related to their atomic complexity. The list of elements continues with berylium, boron, carbon, nitrogen, oxygen, fluorine, neon, etc., up to uranium. Included are familiar elements like copper, iron, silver, and gold. However, many familiar substances are not elements, but are compounds or mixtures. For example, water is a compound of hydrogen and oxygen, and brass is a mixture of copper and zinc.

From the comments above, we can conclude that the chemical and physical properties of matter are determined by its atomic structure. A noteworthy fact is the coincidence between the simplicity of the first two elements, atomically speaking, and their overwhelming 99-plus percent share of all matter. It is customary in astronomical studies of the composition of the stars to refer to "hydrogen, helium, and 'metals' " as the stellar ingredients. "Metals" is commonly understood as anything but the first two elements.

### States of Matter and Temperature

The definitions of solid, liquid and gaseous matter give a clue to the nature of their differences. Solids have their own shape, liquids have the shape of their containers, while gases expand to fill all the available volume of the container.

On the microscopic scale, the differences are directly traceable to the strengths of bonding between atoms. These bonds, which are electrical in nature, are very strong in solids, weaker in liquids, and very weak in gases. The phenomena of melting and vaporization, which are *changes* of states for a material, can be understood as occurring because of a weakening of the atomic bonding, due to an addition of heat energy. The presence of heat in matter manifests itself as a continual vibration or agitation of the atoms constituting it. The amount of agitation is a measure of the temperature of the matter. Hence, adding heat raises the state of agitation and reduces the bonding effect between atoms.

The Kelvin scale of temperature has been invented to utilize this idea. Zero degrees Kelvin ($0°K$) corresponds to the condition of no motion of atoms in a material.* This occurs at approximately 473 Fahrenheit degrees below the freezing point of water. The Kelvin scale is about twice as coarse as the Fahrenheit scale, so one Kelvin degree is about two Fahrenheit degrees. Room temperature ($68°F$) is about $293°K$ above the zero of the Kelvin scale (Figure 3–10).

The material comprising stars is at temperatures which are high enough to overcome the atomic bonds of any liquid or solid; thus, the matter is gaseous. In a gas, the atoms or molecules (bound combinations of atoms) rush around at high speeds colliding with one another in a scene of general chaos. The higher the temperature, the greater the speeds, and the greater the effect of these interatom collisions. At stellar temperatures, the collisions are easily sufficient to affect the structure of the atoms themselves. We will see that light is emitted in the process.

---

* This zero is absolute, because there is no way to define temperature lower.

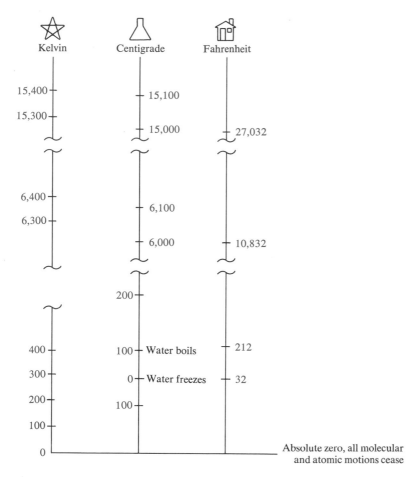

Figure 3–10. *The Kelvin temperature scale is a true physical temperature scale: 0°K is really zero. Note that since the temperatures of stars are measured in thousands of degrees, the few hundred degrees difference between the Kelvin and Centigrade scale is not very important. To get a feeling for large Kelvin temperatures, double them and subtract 10 percent to convert to Fahrenheit.*

## The Structure of Atoms

There are three basic subatomic building blocks of atoms: the *electron,* a very light particle, and two relatively more massive particles, the *proton* and the *neutron,* which each have about 2000 times the electron mass.

While the neutron has no electrical properties, the electron and proton both do. There is, in fact, a force of repulsion between two electrons and also between two protons, while electrons and protons attract each other. These electrical properties are vested in a quantity called *charge* (an attribute distinct from mass). Each charged particle has one unit of charge, by definition.

By convention, the proton's charge is called *positive,* and the electron's charge is *negative.* The adage "like charges attract, unlike charges repel" is applicable. Because charge is distinct from mass, the electrical forces between the parts of an atom are distinct from the gravitational forces between these parts due to their masses. Actually, gravitational forces in atoms are much weaker than electrical forces, and they may be neglected in considering the behavior of the atom's components.

Our picture of an atom's workings is really a construct, or model, put together from various clues, because we cannot see an atom in the usual sense. Decades ago, Niels Bohr introduced a picture of the atom which is still very useful. In the Bohr model, the simplest element, hydrogen, consists of one proton and one electron, with the electron "orbiting" the central proton, as in Figure 3–11. Also shown are helium and lithium, with a more complicated central region, or *nucleus.*

More complex atoms are built up with greater numbers of neutrons and protons, usually in about equal numbers. Every atom attempts to have as many electrons as protons to be electrically neutral. If some event, such as a collision between atoms, removes an electron (from a neutral atom), the atom then has a net positive charge, and will tend to attract any nearby vagrant electron and thus restore electrical neutrality. As atoms become more complex, with more protons and neutrons in the nucleus, there is a requirement for more electrons. It is found that the electrons arrange themselves about the nucleus in a shell-like manner. The first shell may have no more than two electrons; the second, eight, etc. The inner electrons of an atom are much more tightly bound to the nucleus by electrical forces, while the outer electrons are more easily lost by atomic collisions. In some atoms, notably metals, the outer electron is so loosely bound, it is free to roam far and wide. Such materials are very good conductors of electricity.

## Atoms and Energy

If we could isolate any atom of hydrogen, for example, for examination, we would find its electron at a specific distance from the nucleus. An undisturbed atom of hydrogen in this condition is said to be in its *ground state* and its electron is as close to the nucleus as it can get. Since the electron and the nuclear proton are oppositely charged, it follows that

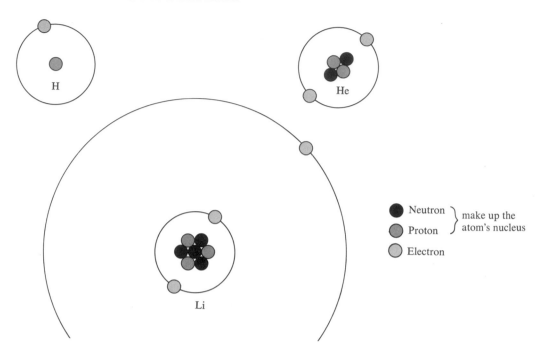

**Figure 3–11.** *The first three elements' atomic structure. Note the building-up of the first shell in helium; lithium's third electron goes into a second shell. Hydrogen's notable chemical activity stems from its incomplete first shell; helium's complete shell assures its inertness; lithium is a metal, and its outer electron is loosely bound.*

the electron is most tightly bound to the atom in this state. However, suppose that we have a container filled with this gas. In this case, the atoms will collide with each other and with the walls of the container. If the gas is heated, the random motions of the gas atoms will increase in speed, and the resulting collisions will be more violent. Collisions which are violent enough can disturb the atomic structure. For example, some of the energy of the collision may be used up to displace the electron from its innermost orbit to one further away.

It is worth emphasizing that an atom must *receive* energy in order for this to happen, because the attractive force of the nucleus must be overcome. Curiously, each type of atom has, in effect, certain allowed distances at which an electron can orbit. It cannot orbit at any but these specific distances. Figure 3–12 pictures a hydrogen atom in which the

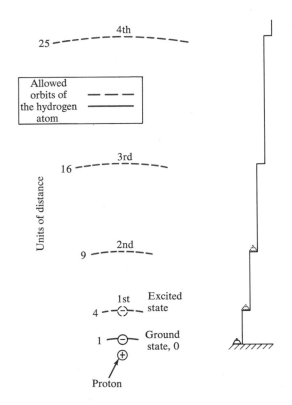

Figure 3–12. *The energy states of the hydrogen atom. Energy is required to "lift" an electron from the ground state, as suggested by the analogy at right. Only one orbit at a time is occupied.*

allowed orbits are drawn to scale. Hydrogen may have no more than its one electron in one of its allowed orbits at one time. When the electron is in the second allowed orbit, the atom is said to be in the *first excited state;* the third level, the *second excited state,* etc. The collisions in a heated batch of gas can raise the atoms to different excited states. Consequently, atoms in excited states are repositories of energy, in this case energy received from collisions. Now, just as it takes a specific amount of energy to lift an object of a given weight to a given height above the ground (working against the force of gravity), it likewise takes a specific amount of energy to "lift" an electron from one of its allowed orbits to another one. Thus we see that although the atom can absorb energy, it can do so only in specific discrete amounts. Collisions are not the only way in which energy may be introduced to an atom; we will discuss another way shortly.

Atoms may also *return* to the ground state by a collision with their neighbors, in a manner resembling the behavior of a cocked mousetrap dropped on the floor. At the impact, the device could uncock, and actually bounce up to a greater height than that from which it was dropped. This would indicate that the stored energy had been converted to energy of motion. Thus, the atoms of a heated gas are not only constantly colliding as a result of their random motions, but they are exchanging energy in the process.

As stated above, the atom is a repository of energy while in one of its excited states. However, atoms have a tendency to spontaneously return to the ground state. In fact, the typical time spent by an atom in an excited state is very short, about $10^{-8}$ seconds. If an excited atom suffers no collisions within $10^{-8}$ seconds after its excitation, it will spontaneously return to the ground state. In this process, the energy reposing in the atom must be eliminated. How is this accomplished? The energy is carried away by *radiation,* in the form of a *single photon.* Let us again consider the diagram of Figure 3–12 showing the allowed orbits of hydrogen. Any possible downward transition will involve a certain amount of energy, which could be carried away by a photon. We may recall that photons of different energies differ in wavelength. The *energy* determines the wavelength of the light, with *larger energies implying shorter wavelengths.* Thus, if we had a heated container of gas, such as hydrogen, we could measure the energies of the transitions between the allowed orbits by noting the wavelengths of the photons which were given off. Since different atoms have different sets of allowed levels, we can actually identify the type of gas which is glowing from the wavelengths of the photons it emits. To do this, it is necessary to separate the light of the glowing gas by wavelength; that is, by color. The light so separated is called a *spectrum,* in the same sense as the word was used earlier in the chapter. Each element then has its spectral fingerprint. The spectra of various heated gasses are shown in Figure 3–13.

Every day one sees countless examples of excited gases producing useful light in the so-called "neon" signs. They are simply tubes of rarified gas through which an electric discharge is passed to excite the gas into glowing.

There are also natural examples of celestial objects with this type of spectrum. One of these is the Orion Nebula (See Figure 3–14 and color plate 13), probably the brightest of the sky objects which glows in this way. It appears faintly visible to the unaided eye as the central "star" in the sword feature of the Orion constellation.

### Atoms Absorb Photons

We have seen that an atom can emit a photon when an electron drops from an upper to a lower state. The photon carries away the energy needed to "lift" the electron. Suppose we place a mirror in the path of the emitted photon and return it to its source. What happens then? The photon meets the atom it has left in a lower state, is absorbed by the atom, and brings us back to our starting point. Atoms can indeed

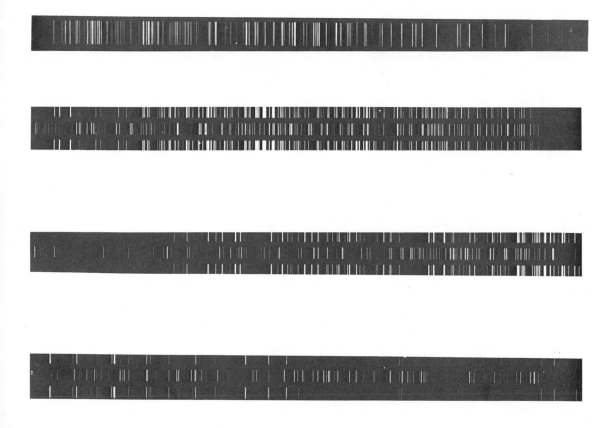

Figure 3–13. *The emission spectra of some heated materials. The top is a mixture of helium and neon gas. The bottom three spectra show iron vapor and neon, flanked by thorium vapor and neon.*

absorb photons as well as emit them. They are not receptive to all photons in this way; they are quite discriminating. Atoms only absorb photons which have energies to lift the electron into an *allowed* orbit for the particular kind of atom. It is easy to demonstrate the effects of this discriminating absorption of light by a gas. Figure 3–15 shows an experiment in which a white light (which has light of all visible wavelengths) falls on a gas and is *selectively* absorbed. When an atom absorbs a photon, it may re-emit one in any direction; it has no memory of the direction of the incident photon. Photons in the incident beam, which are absorbed, are scattered in all directions at the expense of the initial direction.

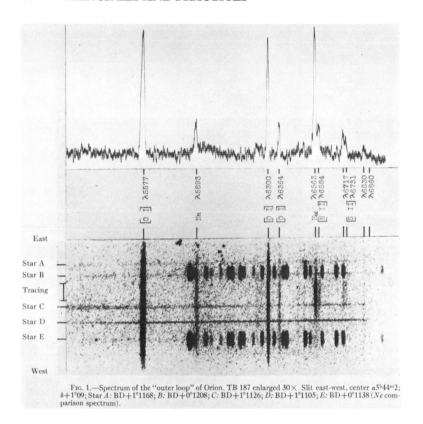

FIG. 1.—Spectrum of the "outer loop" of Orion. TB 187 enlarged 30×. Slit east-west, center α5ʰ44ᵐ2; δ+1°09; Star A: BD+1°1168; B: BD+0°1208; C: BD+1°1126; D: BD+1°1105; E: BD+0°1138 (Ne comparison spectrum).

Figure 3–14. *The spectrum of the "outer loop" of the Orion Nebula (the nebula is shown in the color plates). The nebula has the bright line spectrum characteristic of a tenuous gas. The very low density is evidenced by the appearance of the lines identified by the square brackets, which are never seen in the terrestrial laboratory.*

The wavelengths at which absorption occurs depend on the energies between the allowed levels of the particular atoms of the gas, just as in emission. We note in Figure 3–15 that the spectrum of the scattered light (an emission spectrum) is complementary to the transmitted light (an absorption spectrum). In both instances, then, we are able to make use of the spectral "fingerprint" for identifying the heated material.

Now we can begin to understand the coded message in Figure 3–1. The absorption spectrum of the stars must arise under conditions analogous to Figure 3–15. We can thus picture the situation in the outer

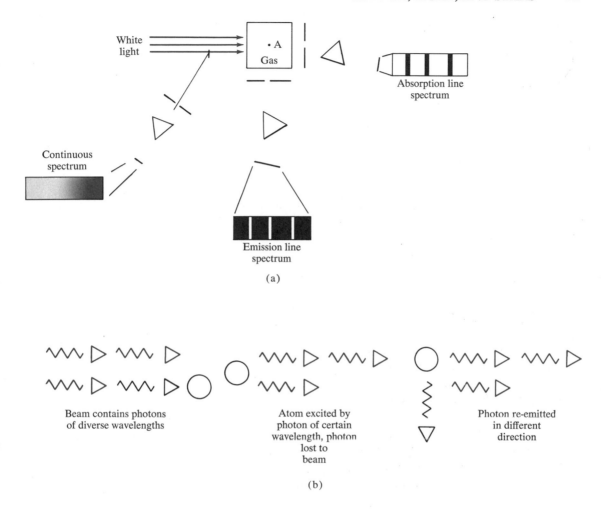

Figure 3–15. *Three types of spectra found in nature: (a) a macroscopic view; (b) a microscopic view.*

layers of stars (in Figure 3–16) as a hot region overlaid by cooler gases. The hot region provides the illumination which is selectively absorbed by the outer layers. Note that the elements which happen to give rise to the absorption features (called *lines*) may be identified. The outer layers of a star (except for the sun) are all that we can study directly. On most stars, this region is only a few hundred miles thick. When compared with the bulk of a star, this is just a relative apple-skin thickness.

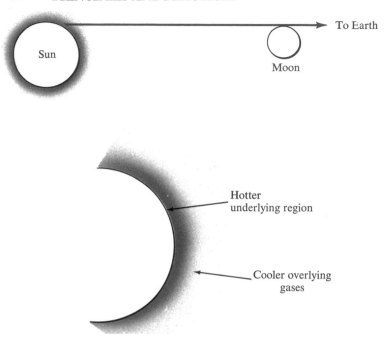

Figure 3–16. *The outer, cooler stellar layers (typically the last few hundred miles) imprint the absorption features on a stellar spectrum.*

The above picture of a star's outer layers may be tested very nicely for the sun, because immediately before and after the total phase of a solar eclipse, we see the sun's outer layers *only,* and the moon blocks the fierce light of the disc. Then we do see an emission spectrum and it is the complement of the absorption spectrum.

### Ionization of Atoms

When the temperature of a gas rises high enough, an atom's outer electron, or electrons, may become completely dislodged, a process termed *ionization.* Since the different species of atoms differ in the strength of their grip on their electrons, ionization occurs at different temperatures for different atoms. Further, atoms with more than one electron can be multiply ionized with increasing temperature. Of course, when an electron is removed from an atom, there can be no more absorptions involving its transitions. Let us illustrate temperature dependence of ionization as it is seen in stars. In the sun we see a spectrum rich in absorption lines due to un-ionized (or "neutral") metals. We recall that metals do not have much hold on their outer electrons; thus it is not at all surprising to notice that in stars slightly hotter than the sun (the sun's atmospheric temperature is about 5,800° K), the lines due to absorption by the "neutral" metals are beginning to fade. This is true because a greater fraction of the metals are in the ionized state. We can see that the absorption lines in the spectrum of a star can give clues to the temperature of its atmosphere. This will be discussed further in Chapter 6.

## The Spectrum of an Incandescent Object

We have seen that the temperature of a star could be inferred, in principle, from the absorption lines in its spectrum. The rest of the spectrum apart from the absorption lines is called the *continuous spectrum*. This too contains clues to the temperature of the outer stellar layers.

At the turn of the century, physicists were wrestling with the problem of how to predict the spectrum of a heated object, like a poker in a fire. The evidence of temperature's effect was easily seen as the object was heated: it first glowed dull red, then cherry-red, orange, orange-yellow, yellow-white in succession as its temperature rose. In the laboratory situation, the heated object was arranged in such a manner to be sure that the radiation emitted had interacted thoroughly with the heated body. Figure 3–17(a) shows how this was done. Such a

Figure 3–17. *Illustrating the similarity of the radiation from a perfect radiator (a) to that in the outer layers of a star (b).*

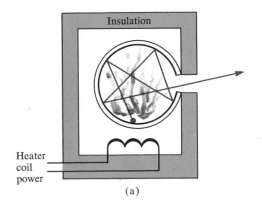

Insulation

Heater
coil
power

(a)

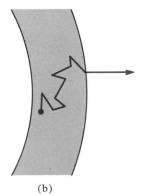

(b)

body is variously called an *ideal radiator,* a *perfect radiator,* or a *blackbody.* The rather odd last term relates to the fact that perfect radiators are perfect absorbers. For instance, any photon entering the observing aperture to the cavity will very likely not get out. The early physicists had to explain the following observed facts: the amount of light and the spectrum of the light from a perfect radiator did *not* depend on the composition of the body, but *only on its temperature;* the total amount of radiant energy of all wavelengths given off by a unit of area was proportional to the fourth power of the temperature (measured in Kelvin degrees); and the *wavelength* at which most of the energy was emitted was inversely proportional to the temperature. It may be surprising to learn that the stumbling block to an explanation was that the discrete, or photonlike nature of light had not yet been recognized. Max Planck, in 1901, incorporated this into a new and correct theory of radiation from a blackbody. Planck's law gives the actual amount of energy at each wavelength in the spectrum of a blackbody at various temperatures, as shown in Figure 3–18. As the temperature is raised we note the following: the amount of energy given off at each wavelength is higher; the total energy, summed up for all wavelengths, is larger; and the wavelength at which the peak of the curve occurs is shorter for the higher temperatures.

The question arises as to whether Planck's law could be used to determine a temperature for the outer layers of a star. If stars were ideal radiators there would be no question. However, we can note that, at least at first glance, a star would seem to bear remote resemblance to the radiating object of Figure 3–17(a), which was turned in on itself. Actually, the outer layers of a star are semitransparent, and photons which are emitted at points in these layers rarely escape to the outside without being absorbed and re-emitted many times. This process closely resembles what happens to a photon emitted from the interior of an ideal radiator. Figure 3–17(b) shows these processes for a star and blackbody, emphasizing their similarity. In both cases, photons have difficulty getting out. Under these circumstances, we can expect that the continuous spectrum of a star resembles a blackbody; therefore, we could tell something about its temperature. In Figure 3–19 we see that the spectrum of the sun tolerably well matches that of an ideal radiator at 6,000° K. We can also see that the sun is not exactly an ideal radiator.

Thus we have another tool at our disposal for finding the temperatures of stars. In this method, we assume that the temperature of the star equals that for the Planck curve which best matches the observed continuous spectrum. The mathematics giving the form of the curve depends only on temperature, but the precise formula need not concern us here.

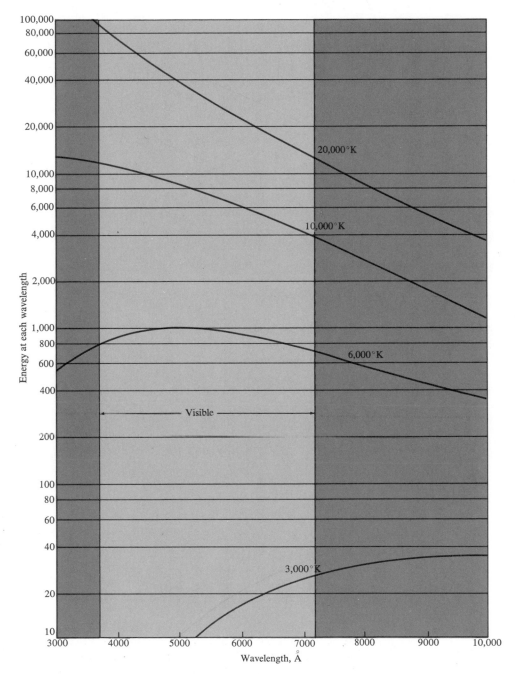

Figure 3–18. *The spectral distribution of the energy of an incandescent object which is a perfect radiator. The amount and type of radiation is governed solely by the temperature.*

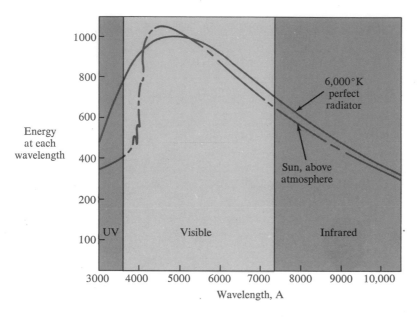

Figure 3–19. *The spectrum of the sun's radiation outside of the earth's atmosphere closely resembles that of a perfect radiator at 6000°K. Note the sun's relative lack of ultraviolet light.*

## Starlight and the Doppler Effect

The Doppler effect in light is more conveniently discussed in terms of wavelength changes rather than frequency changes. Einstein showed that the speed of light must be a constant; so, because of the general formula relating wavelength and frequency, we will always have for light:

wavelength $\times$ frequency = a fixed number, about 299,997 km/sec

Therefore, if there is a change in frequency, for instance an increase, there is a corresponding decrease in wavelength for the light, and vice versa. Suppose, for example, we were to approach a red light source. Its frequency would increase, or, alternately, its wavelength would decrease. There is then in principle a color shift in the light moving it towards the blue end of the spectrum, that is, towards the shorter waves. The *amount* of the shift depends on the amount of the speed between the source and observer, compared to the speed of light. These shifts are usually very small in practice, because most familiar speeds are much smaller than the speed of light. (Thus, it would not help one to claim in traffic court that a red light was actually green in appearance as it was approached, due to the Doppler effect. One could then stand accused of speeding at a considerable fraction of the speed of light!) The only objects in nature for which color changes can be noticed are the distant and very rapidly speeding galaxies (to be discussed later).

The operation of the Doppler effect may be evident when we examine the *spectrum* of a star. Let us illustrate this statement in Figure 3–20, where we have reproduced the effect which would be seen in the spectrum of the sun if it were in motion at 100 km/sec with respect to earth. The sun's current speed relative to earth is, of course, almost exactly zero. A 100 km/sec speed for a star relative to the earth would not be an unreasonable value. We see that the clue to the star's motion is not the change in the star's color, which would not even be per-

Figure 3–20. *The effects on the sun's spectrum of a hypothetical 100 km/sec speed away from the earth (top), and toward the earth (bottom). Speeds along the line of sight can be determined to a precision of about 1 km/sec under the best circumstances.*

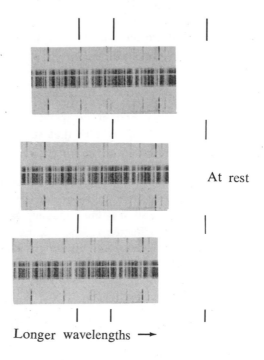

At rest

Longer wavelengths ⟶

ceptible, but rather a slight shift in the position of the star's spectrum. This shift is detectable by the positional change of the dark lines, compared to their position when the source is at rest. The amount of the shift is proportional to the speed of the source, while the direction of the shift depends on whether the source is approaching or receding. Figure 3–20 show how the shift in the position results in new wavelengths for the dark lines. Now choose any line and calculate: new wavelength minus old wavelength and divide the result by the old wavelength. Using the symbol $\lambda$ for wavelength:

$$\frac{\lambda_{\text{shifted}} - \lambda_{\text{rest}}}{\lambda_{\text{rest}}}$$

This number is a fraction, the fraction of the speed of light that the source is traveling.* An example of this calculation has been made in Figure 3–20, in which the center spectrum is from the sun at rest and the top and bottom spectra represent either the sun, or a sunlike star having a speed of 100 km/sec of approach and recession, respectively. In actual practice, the speeds of stars are reckoned from shifts in their spectra, which are recognized by comparison with the spectrum of a gas or special arc lamp. These produce many bright lines whose wavelength positions are very accurately known. Star speeds can be determined today to a precision of about 1 km/sec, which corresponds to a very small shift in a star's spectral lines. A special measuring microscope is used to measure the shift.

What we have here is an astronomical speedometer, which may be used to investigate the speeds of the star from clues in the spectra of their light. Neither tape measure nor stopwatch is necessary, nor do we have to leave the earth to carry out the measurements. We must realize, however, that this is only a one-dimensional picture of a star's motion. We only learn of a star's speed of recession or of approach, that is, along the line of sight to that star. This speed is known as a *radial velocity*. As Figure 3–21 shows, a given star's motion may be such as to move it *across* the line of sight as well, in other words, across the sky. Because of the great distance of the stars, these motions become apparent only after careful measurement on photographs taken decades apart. This motion is known as a tangential motion and will be further discussed in Chapter 7.

The determination of radial velocities plays a major role in our knowledge of the masses of the stars, their organization in the Milky Way Galaxy (the great system of stars to which we belong), and even

---

*Doppler's formula is: $\dfrac{\lambda_{\text{shifted}} - \lambda_{\text{rest}}}{\lambda_{\text{rest}}} = \dfrac{v}{c}$

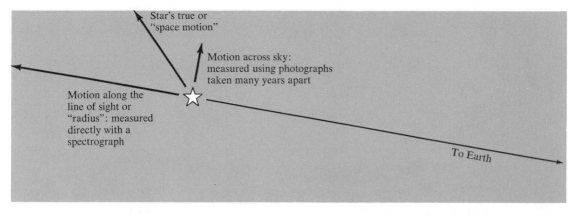

Figure 3–21. *A star's true or "space" motion, here resolved into two components. The two components are measured in very different fashion. Motion along the line of sight is known as a "radial velocity," while motion across the line of sight is a "transverse velocity" (usually measured in arc-secs/year, and called the star's "proper motion").*

of the evolution of the universe itself. Yet, this is only a fraction of the information conveyed by a spectrogram. In a sense, this information is only incidental to its telling us something about conditions in the matter in which it originates.

## Questions

1.  Give an example of light's wavelike and its particlelike properties.

2.  What determines the energy of a photon? Why is the ultraviolet component of sunlight the most energetic?

3.  What are the three states of matter? Which is most prevalent in the universe?

4.  What is temperature, on an atomic scale?

5.  How is the structure of the atom related to its ability to store energy?

6.  What is ionization?

7.  Show how an absorption spectrum, an emission spectrum, and a continuous spectrum are formed.

8.  Give an example of the three types of spectra as they occur in nature.

# 4 Modern Observational Astronomy

## Introduction

The aim of this chapter is to give some insight into the current state of observational astronomy. The importance of observations in this science was once well expressed by a famous astronomer who said that if we hadn't seen the stars, theoreticians might have long ago concluded that they could not exist. Of course the remark was made in jest, since he was a theoretician himself. It is really not much of an overstatement. Throughout the history of astronomy, observations have led the way to progress. We might recall Brahe and Kepler as two of the earliest examples of this process. And today, for example, we are faced with the observations of the *quasars* which pose quite a dilemma to current theoreticians. There certainly are times, however, when observers are working to verify some prediction of theory. For example, the discovery of the *pulsars* verified the existence of the predicted neutron stars.

The technology of recent years has had a profound effect on observational work. Whole new branches of observational astronomy have been created, such as radio, infrared, ultraviolet, and X-ray studies. Some of this work can be accomplished only by the use of above-atmosphere stations, provided by the use of balloons, rockets, and satellites.

Direct study of the moon, planets, comets, and the interplanetary medium have been initiated by space probes. Important contributions have also been made by the Skylab manned satellite.

The effect of technology on ground-based optical astronomy has likewise been great. The data-gathering equipment at observatories can be made more efficient and can be interfaced with a computer, allowing the observer to analyze the data as it is received. The telescope is regarded more and more as just one element in a complex data system.

Let us first learn how telescopes work, where they are located, and something about the different types of observations for which they may be used. Following this, we will try to give an impression of what it is like to actually perform observational work at the telescope. Finally, we shall examine some of the new fields of observational work.

## The Telescope

The first observational tool used in investigating the sky was the eye, but when the telescope came along, it vastly increased the region of the heavens available for study. At first, the telescope served as an aid to the eye, but with the advent of photography, it gradually assumed the role of a camera. With the advance of technology, it has become an analytical instrument helping to reveal the nature, composition, and evolution of the stars.

Anyone who has observed the sky with telescopic aid (binoculars will actually do nicely for an overall perusal) would have no difficulty in discovering how useful telescopes are in astronomical observations. First of all, stars fainter than those detectable by the unaided eye are made visible. On a very clear night, with no moon, and away from the city, some three thousand stars may be seen over the whole sky with the unaided eye. (This figure may seem improbably small, but try counting the stars in a small area, and estimate from that the number over the whole sky). With the aid of typical binoculars, the number of visible stars approaches 25,000.

With a 6-inch telescope, a size used by many amateur observers, this number rises steeply to 300,000. Imagine the number of stars which are available for study with a telescope such as the 200-inch instrument at Palomar Mountain. Obviously, not every star in the sky could be studied by one or even a battery of very large telescopes.

Stars invisible to the unaided eye become visible under telescopic aid because the opening, or *aperture,* of a telescope is considerably larger than that of the eye. A larger aperture means that more light from a star can be collected, resulting in a brighter image. Figure 4–1 shows parts and functions of a telescope which uses lenses to focus the light. Known as a *refracting* telescope, it was the type first used astronomically by Galileo (Figure 4–2). Stars may be regarded as geometrical points of light at great distances. Rays of light leaving a star, which reach the eye or optical instrument, are traveling in parallel paths, because, as seen from the star, the "top" or "bottom" of the opening of the receiving instrument is in virtually the same direction. The rays of light collected by the large lens, or *objective,* form an image which may be conveniently ex-

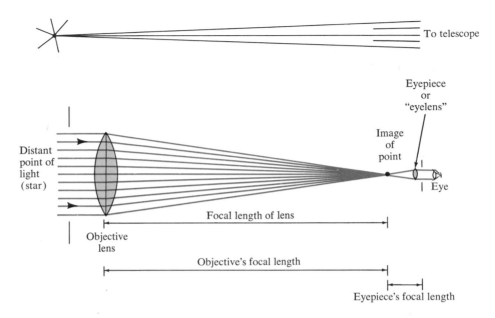

To telescope

Eyepiece
or
"eyelens"

Image
of
point

Eye

Distant
point of
light
(star)

Objective
lens

Focal length of lens

Objective's focal length

Eyepiece's focal length

Figure 4–1. *Light rays from a point source a great distance away (e.g., a star) will be parallel at the telescope (top). The first optical element, whether lens or mirror, brings the light rays to one point, the focus. Here, with the help of a magnifier (or eyepiece), the image may be examined by the eye. Note that the cylinder of light filling the eye is equal to the area of the telescope, whereas the unaided eye accepts light only through its small (7 mm) pupil. It is thus obvious why the stars appear brighter in a telescope. The focal length of the eyepiece is important in fixing the telescope's magnification, which is the ratio of the focal length of the objective divided by that of the eyepiece.*

amined by the eye with the aid of a magnifier, called an *eyepiece*. Observe that these two optical elements working together have reduced the large cylinder of light at the front of the instrument to one of just the right size to enter the eye; no light is wasted. This gain in the amount of light entering the eye by use of a telescope is directly responsible for the brighter image of a star viewed with telescopic aid. This is why the astronomer is prompted to build larger and larger instruments. The larger the instrument, the fainter the stars which can be seen, or

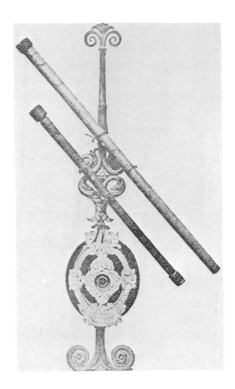

Figure 4–2. *Galileo's telescopes, the first astronomical refractors. The largest had an aperture of 1-inch (about 2.5 cm) and magnified about three times.*

photographed with a given time exposure. Since the area of a circle is proportional to the square of its radius, doubling the diameter of a telescope's objective increases the area by four times and four times as much light may be collected from celestial objects. Figure 4–3 shows a portion of the sky as photographed by two telescopes of different size, but with equal exposure times.

Let us now consider how a telescope produces images of nonstellar objects such as the sun, moon, planets, or galaxies. The manner by which the objective forms the images is indicated in figure 4–4(a). Unlike the case for stars, the brightness of extended objects is not aided by increasing, for example doubling, the scale of a telescope.

Utilizing the example above, imagine a telescope doubled in scale —four times more light will be admitted by its aperture (Figure 4–4). On the other hand, the focus of the larger instrument is twice as far from its objective. This means that the image will be four times larger than with the smaller instrument, so the light will be spread over a four times larger area. Four times the light spread over four times the area results in an image of the same brightness. If the two telescopes photographed the moon, the same exposure time would be used; the larger telescope merely produces a *larger* image.

Figure 4–5 shows the largest refractor in the world, the 40-inch

Figure 4–3. *Two photos of the same region of the sky as taken with two telescopes, one having twice the aperture of the other.*

Yerkes telescope. Its objective, like that of all astronomical refractors, is actually composed of two optical elements working together. This is necessary because glass refracts, or bends, light of different colors in slightly different directions, a process known as *dispersion*. The dispersion occurring in the first element is corrected in the second, as shown in Figure 4–6.

Most modern telescopes are *reflectors* (which utilize mirrors) for a number of reasons. In a refractor, the light must traverse the glass elements, so this glass must be highly uniform throughout; no bubbles, pits, or other blemishes anywhere in the glass can be tolerated. On the other hand, glass discs for mirrors need only have a decent *surface*. Furthermore, modern astronomy requires large instruments, for which the refractor is ill-suited. While mirrors may be supported over their entire back surfaces, lenses can be supported only at the edges. Long ago, it was found that the largest size possible for a refracting objective is 40 inches, if it is not to sag under its own weight. On the other hand, the size of the largest reflecting mirror which could now be built is estimated at 400 inches, but even 600 inches is considered feasible by some experts in telescope design. The area of this latter giant would exceed the area of the 200-inch by $3^2$, or nine times. The largest reflector in the world is currently the 236-inch reflector in the Soviet Union. Reflectors, besides permitting large size, are also free of dispersion, since there is no traversal of glass by the rays of light. A third advantage of reflecting optics is that they may be used where glass might not have sufficient transparency to the light being studied. There is also quite a savings in size made possible by the use of a reflector rather than a refractor, as will be seen.

Figure 4–4. *(a) The image of an* extended object *such as a planet (versus a star) may be regarded as a group of contiguous image points (but see the section on diffraction). Each object point thus acts like a star. (b) A luminous extended object, such as a planet, will have the same brightness in all telescopes having the same ratio of focal length to diameter. The scale of these telescopes is 2:1; the smaller instrument accepts only one-fourth the light of the larger but distributes it over only one-fourth the area of the larger instrument.*

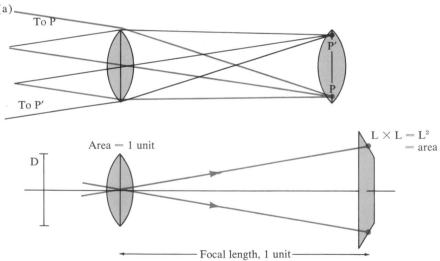

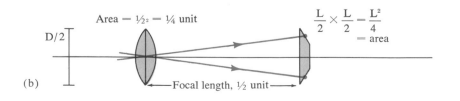

Figure 4–5. *The 40-inch (102-cm) telescope of the Yerkes Observatory, the world's largest refractor.*

The manner in which the mirror brings light to a focus is illustrated in Figure 4–7. Unlike the refractor, the reflecting instrument forms its image back in the direction of the incoming light. With a smaller instrument, especially, it would be impossible to get at the image without blocking a considerable amount of the incoming light. Newton, who built the first reflecting telescope, had the idea to place a small flat mirror to deflect the rays to the side, where the image would be examined without the observer being in the way of the incoming beam, as shown in Figure 4–8(a). Another solution to the problem, which also had the advantage of shortening the telescope tube, was later devised by the Frenchman Cassegrain. He boldly drilled a hole in the principle mirror, directing the rays through it by means of a curved second mirror, shown in Figure 4–8(b). The Cassegrain telescope is actually about three times shorter than an equivalent refractor. At the present time, the simplicity of the Newtonian telescope has endeared it to many amateur observers, while the majority of observatory instruments tend to follow Cassegrain's model.

Many of the large instruments are adaptable to both focus arrangements, as well as having a focus fixed in position—the Coudé focus shown in Figure 4–8(c). Extra mirrors are required in a Coudé arrangement to keep the telescope's focus stationary, regardless of the point on the sky at which it is trained.

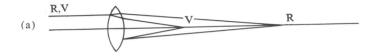
(a)

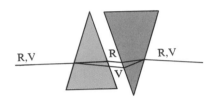

(b)
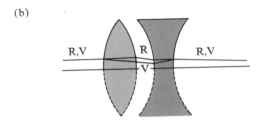

Figure 4–6. *(a) The prismatic behavior of a simple lens. (b) A second lens may be utilized to recombine the colors, a principle utilized in the "achromatic doublet" lens. R: red rays; V: violet rays.*

### Observatories around the World

Scattered around the globe are scores of professional observatories at which, on clear nights, observers will be probing the skies. Almost all of these installations belong either to universities or to governments. The number of observatories with instruments of about 80 inches or larger is quite small. These are listed in Table 4–1. Something important to note is that the worldwide distribution of large telescopes strongly favors the Northern Hemisphere. The large southern telescopes are, for the most part, relatively new; and since the southern sky cannot be observed with northern instruments, there are many mysteries of the southern sky yet to be revealed. Even today, it is regarded somewhat as a *terra incognita* by the astronomical community. Unique to the southern sky, for example, are the Magellanic clouds, miniature galaxies and satellites of our own galaxy, and our nearest galactic neighbors (Figure 11–4). In addition, the center of our own galaxy, the Milky Way, passes overhead in good observing position.

## TABLE 4–1

*The World's Great Telescopes*

| Location | Size in inches (meters) |
|---|---|
| Pestukov, USSR* | 236 (5.9) |
| Hale Observatories, California | 200 (5.0), 100 (2.5) |
| Kitt Peak National Observatory, Arizona | 158 (4.0), 84 (2.1) |
| Cerro Tololo Interamerican Observatory, Chile (Kitt Peak Southern Station) | 158 (4.0) |
| Anglo-Australian Telescope, Australia | 150 (3.8) |
| European Southern Observatory, Chile | 150 (3.8) |
| Hawaii, France, Canada Joint Telescope, Hawaii* | 150 (3.8) |
| Lick Observatory, California | 120 (3.0) |
| McDonald Observatory, Texas | 107 (2.7), 82 (2.1) |
| Crimean Astrophysical Observatory, USSR | 102 (2.5) |
| Hale Observatories Southern Station, Chile* | 100 (2.5) |
| Royal Greenwich Observatory, England | 98 (2.5) |
| Steward Observatory, Arizona | 90 (2.3) |
| Mauana Kea Observatory, Hawaii | 88 (2.2) |

* Not yet in operation.

Observing time on a large instrument is very precious, since there are so few instruments of very large size, and the number of astronomers is not small. Even though international cooperation among observatories is very good and some time is available for guest investigators from smaller astronomical centers, there is not nearly enough guest time to satisfy the needs of the astronomical community. In the United States, the response to this need was the creation of the National Observatory at Kitt Peak, operated by the Associated Universities for Research in Astronomy in cooperation with the National Science Foundation. Visiting astronomers and graduate students account for the bulk of observing time using its extensive facilities. The rest of the time is utilized by the Kitt Peak permanent staff, as well as for testing purposes. Applications for time are submitted several months in advance and the scientific program is scrutinized by a panel to insure the most efficient use of the instrumentation. The quite extensive facilities include, in addition to the two reflectors listed in Table 4–1, a 50-inch experimental telescope, two 36-inch and four 16-inch telescopes, and a number of smaller specialized instruments. The mountain observatory also features the very impressive McMath Solar Telescope and the three telescopes of

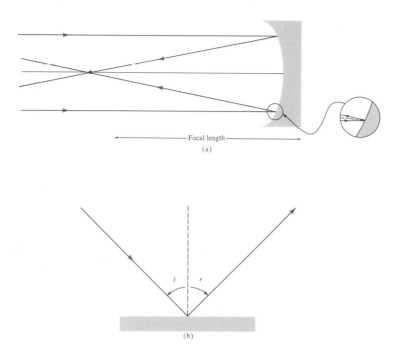

Figure 4–7. *(a) Image formation by a concave mirror; the object is a distant point source of light. (b) The geometry of reflection: the angle of the ray incident equals the reflected angle, or* i = r.

the Steward Observatory. Kitt Peak also has a southern station, the Cerro Tololo Interamerican Observatory on La Serena Peak in Chile, which has very extensive facilities.

### Telescope Sites

The optimal placement of an observatory is a matter of considerable concern and study. After all, a large instrument can represent an investment of millions of dollars, so one wants the device to be used on as many nights as possible. The ideal site is high and dark, and has an unusually large percentage of clear, dry nights. A very dark sky is especially necessary for the largest telescopes because they can reach the faintest-appearing stars, which must not be overwhelmed by skylight. The high altitude means that the instrument will be above a good percentage of the atmosphere, thus reducing the effects of image blur and atmospheric absorption of the light. Dry weather (low humidity) is

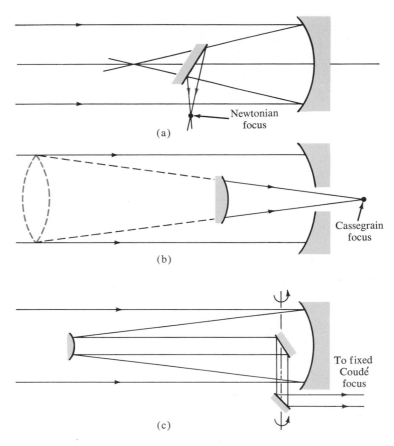

Figure 4–8. *Three methods of bringing the focus to a point convenient for observation or measurement: (a) the Newtonian focus, (b) Cassegrain's solution, proposed later. In the Newtonian, the secondary mirror is flat, but it is convex in the Cassegrain. In (c), extra mirrors allow the Coudé focus to remain fixed while the telescope follows the stars.*

necessary for observations in the infrared region of the spectrum, due to the absorptive properties of water vapor. Usually, these considerations mean that observatories are located on mountain peaks in regions where desert-like climates exist.

Potential sites are first thoroughly tested for clarity and image steadiness, as well as number of clear nights per year. Such a site survey may take years. Pollution of the air, both natural and man-made, is

Figure 4–9. (a) The dome of the 158-inch (4-meter) Nicholas U. Mayall reflector of Kitt Peak National Observatory, located forty miles southwest of Tucson, Arizona (top). The immense size of the instrument can be appreciated by noting the size of the door on the catwalk around the dome, or by noting the relative size of the 90-inch (2.2-meter) dome, at its base, left. Several instruments of lesser size are also visible. The strange instrument in the far left portion of the photo is a solar tower telescope. Right: the Mayall telescope pointing northward. The mirror is located at the center of the "horseshoe," lower right. This telescope, like other very large instruments, permits observing at the prime focus (the first one) by means of an observer riding in the cage at upper left. This whole unit may be flipped 180° to bring the secondary mirror onto view.

also an important consideration, because dirty air scatters and absorbs the light. Astronomers must also worry about "light pollution" from urban areas. A large city may effectively put a damper on certain types of important astrophysical measurements, even though it may be miles away. For example, in California, the Mt. Wilson 100-inch is seriously hampered by the lights of nearby Los Angeles, which brighten up the whole sky.

Even the great 200-inch suffers somewhat, with a real threat of very serious problems ahead. An illustration of the problem is the spectrogram of Figure 4–10. It is sobering to think that the light which exposed that spectrogram traveled for billions of years from its distant source, and that it was contaminated in the last microsecond or so of its existence. Some urban centers have recognized the problem of their nearby observatories and responded with useful ordinances. Recently,

Figure 4–10. *The quasar 3C295 in the constellation of Bootes (top), and its spectrogram (below). The bright lines top and bottom in the spectrogram are from the comparison light source in the instrument. The faint bright lines in the center spectrum are due to the night sky. The very bright line comes from mercury vapor lights in the Los Angeles area. The faint horizontal line just below center is the quasar spectrum.*

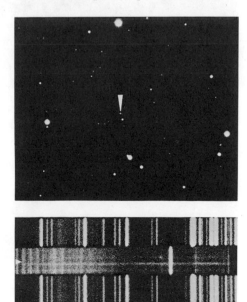

the city of Tucson, Arizona, ordered top covers on all outdoor city lights, to minimize interference with nearby Kitt Peak.

A good example of an excellent mountaintop site is the Mauana Kea Mountain site of the University of Hawaii's observatory. Here is located the highest large telescope in the world, at an altitude of 13,000 feet. It also enjoys the clear, clean air of the Pacific, free of continental

Figure 4–11. *The dome of the 2.2-meter telescope at Mauana Kea, Hawaii. The snow is not unusual, as the altitude is 13,000 feet (3600 meters). Here the visibility and transparency are exceptionally fine, perhaps the world's best. A careful search reveals the domes of the 1.5-meter Air Force, and Planetary Patrol telescopes.*

smog. Despite its island location, the altitude results in a climate exceeding our western deserts in aridity.

Many astronomers agree that the finest all-around observing location in the world is found at Cerro Tololo, at 8,000 feet on La Serena peak in Chile. The climate of northern Chile is very favorable to astronomy, being desert from its west coast to its mountain range. The region boasts some 330 nights of crystal clear skies yearly. Occasionally, star images of about 1/4 arc-sec are seen on nights of very low turbulence. At most sites, 1 arc-sec star images are rare and considered excellent.

One new observatory was actually chosen as a result of satellite observations, a useful feedback of space astronomy to the classical branch. Photos of the North American continent showed that the region of the central Baja Peninsula was persistently clear throughout the year. The region is, of course, a desert, remote and uninhabitable, and thus a perfect place for astronomers. Before long, an interamerican group from Mexico and the United States had located a telescope there.

Certain sites have unique advantages. For example, the Lund Observatory in Sweden, due to its high latitude, can take advantage of its almost twenty-four-hour winter nights to observe stars almost continuously, an impossible task for lower latitude observatories.

Some consideration has been given lately to the special problems encountered by the solar astronomer in daytime observations. The main problem is due to the sun's warming the ground and setting up convection, the circulation of warm and cold air, which produces poor seeing. Since it is well-known that water requires more heat energy to raise its temperature a given amount, compared to the land, it seemed that daytime seeing problems would be moderated for an observatory located on the water. The experience at the Big Bear Lake solar observatory of the California Institute of Technology seems to confirm this reasoning, and superb pictures of the sun's disc have been obtained. Figure 4–12 shows an example.

## The Telescope as a Camera

Astronomical photography is basically a simple process. The telescope serves as the camera, and the photographic film or a glass plate with photographic emulsion on its surface ("a plate") is placed directly at the focus as in Figure 4–13. Although glass plates are considerably more expensive and more delicate than film, the dimensional stability which glass provides in the photograph is often very necessary for scientific purposes. One of the principal benefits of photography versus direct visual observation is that the exposure on a photographic plate builds up over many hours, if needed, while the eye's effective exposure time

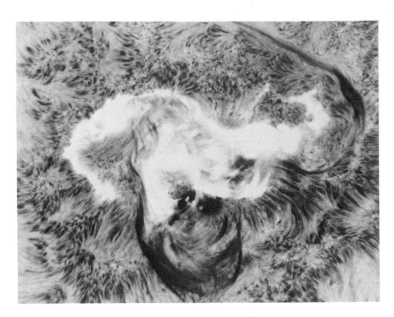

Figure 4–12. *An exceptionally detailed ground-based photo of the sun, showing a close-up of a sunspot and active area. The observatory is located in the middle of a mountain lake in California. Since water heats up slower than the ground, there is less mixing of hot and cold air, which results in superior seeing conditions.*

is about 1/10 second. Many of the photographs throughout this book, especially those of the galaxies, commonly result from exposures of hours on large telescopes. Again, contrasted with visual observations, photography provides a record which may be measured, correlated, and analyzed at any time after the original observation. For this reason, it has been many years since visual observations occupied the major fraction of time on large telescopes.

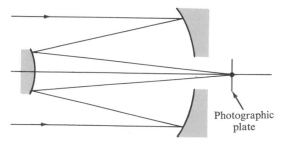

Photographic plate

Figure 4–13. *A Cassegrain telescope used as a camera. The photographic plate is placed at the focus of the instrument.*

Direct photography is used today to determine the position and brightness of planets, stars, and galaxies. Exposures in different colors are used, giving measures of the colors of objects. Occasionally, full-color photographs may be useful in the study of certain objects, and provide aesthetic pleasure as well, as attested to by the color plates in this book.

The most ambitious photographic project ever undertaken was the National Geographic Society-Palomar Observatory Sky Survey. Its purpose was to photograph the whole sky visible from the Northern Hemisphere, revealing stars at least 100,000 times fainter than visible to the unaided eye. Special thought was given to the type of instrument best suited for this job. It had to have a large aperture to gather light effectively, and keep exposure time well below an hour. Further, a large portion of the sky had to be photographed at once. The instrument used is known as a Schmidt telescope, after its developer, Bernhard Schmidt. It can view an area of sky 6 degrees square, equivalent to over 150 moons side-by-side. Even so, over 1,500 exposures were needed to cover the whole sky visible from Mount Palomar. The result of the work is an atlas, which contains photos taken in blue and red light for a total of about 3,000 exposures. Figure 9–5 is a photo looking in the direction of the constellation of Sagittarius, showing interstellar gas and an almost solid view of stars. Copies of the atlas are now widespread at observatories in the United States and throughout the world. They have proved to be an indispensable tool in astronomical research. For example, when radio waves are detected coming from some particular direction of the sky, the atlas can be consulted for a possible optical identification. Furthermore, the countless groups of faint galaxies which were revealed have become the subject of diligent study. The distribution of these groups over the sky itself has become a study with most crucial ramifications for the evolution of the universe. One of the more amusing and serendipitous uses of the plates was the accidental discovery of two very faint, but quite nearby, galaxies, by an astronomer who just happened to be perusing a plate for idle curiosity!

### The Telescope as a Light Meter

We have seen that photographs can be used to gauge the apparent brightness of the stars. Under favorable conditions, an accuracy of a bit under 5 percent in reckoning the brightness of a star is possible. For many purposes, this is not sufficiently accurate; instead, results closer to 0.5 percent accuracy are needed. This occurs in the study of stars which vary in light, for the mapping of light distribution over planets and galaxies, for the study of brightness of stars in clusters, and in many other instances. Such measurements are accomplished with the use of auxiliary instruments, known as *photoelectric photometers,* attached to telescopes. These are sophisticated versions of the light meters found on conventional cameras. Figure 4–14 shows a diagram of a photometer. The heart of the device is a special vacuum tube, called a *phototube,* in which light is converted to an electronic signal, whose strength is a measure of the light striking the tube. By means of a small hole, light from the star and from a bit of surrounding sky is allowed to fall on the tube, and the signal is recorded. Momentarily, the telescope is shifted slightly and the light from the sky alone is noted. By subtracting the second reading from the first, the light of the star alone is obtained. (Stars of constant light may be compared in this manner.) Stars whose light is variable are commonly studied by comparing them to nearby stars known to be of constant brightness.

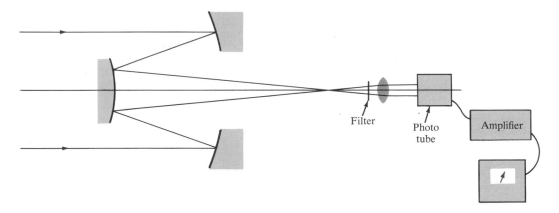

Filter

Photo tube

Amplifier

Figure 4–14. *A Cassegrain telescope used as a sensitive light meter. Light from a star, or other celestial source, is allowed to fall onto the surface of a light detector, which converts it to an electrical signal which may be re-corded, observed on a meter, etc.*

Photoelectric photometry tends to be the most common task of telescopes of moderate size, typically available at the observatories of universities. Studies of apparent brightness and variability in the bright-ness of planets, stars, and galaxies yields important information. Mea-surements of the brightness of these objects as seen through various filters can reveal temperatures and clues to the chemical compositions of the objects studied. This type of research has shown that many stars and galaxies are variable in light, and this variability provides another piece of information in defining the nature of the objects. We shall see in

Chapter 11 that the time scale of variability of the quasars can give an idea of their size, even without knowing their distances. Studies of the light variability of stars have helped determine that certain of them pulsate in regular fashion, alternately swelling and contracting. Also widely studied are the *eclipsing variables*. These are close pairs of stars which eclipse each other as seen from the earth's direction, as they mutually revolve about each other. The stars may be so close that the facing sides are raised to extra-high temperatures by mutual heating, and their spherical shapes become markedly distorted by mutual gravity. Although these stars are too close to be seen as double in the telescope, their periodic light changes, during the course of a revolution, may be analyzed to obtain a surprisingly complete idea of their appearance.

### The Telescope as an Analytical Instrument

Our knowledge of the nature of celestial objects is, of course, due to the light we receive from them. Detailed analysis of this light is therefore the means by which we may learn the most about the source of light. The means by which this analysis is conducted is known as *spectroscopy* and has been with us for many years, at least in its primitive form. As noted, Newton first subjected a beam of sunlight to the dispersive action of a glass prism. The incoming beam of white light was converted into a multihued-beam exiting the prism, which contained all the colors of the spectrum. The first real astronomical use of spectroscopy was an experiment conducted by the Russian Kirchhoff. By using a more sophisticated apparatus than Newton, Kirchhoff first noted the absorption lines in the spectrum of sunlight. Since a German named Fraunhofer did a very thorough job of studying these absorption lines, the sun's absorption features are still called *Fraunhofer lines*. By the turn of this century, it had become possible to record the spectra of many bright stars. They also showed the similar phenomenon of the dark lines, with some stars having a remarkable resemblance to the sun's spectrum, but others being significantly different. The significance of this will be studied in Chapter 6.

Let us have a look at the basic parts of a spectrograph as used on a telescope, in Figure 4–15. The basic principle of operation is that used by Newton, but with a few more optical elements.

The most vital role of the larger telescopes has traditionally been a program of spectroscopic observations. In a rough way, we can see the importance of a large telescope for this work by comparing the photography of a star with the photographic recording of its spectrum. In the latter case, all the colors are spread over the plate, rather than being concentrated in one tiny point of starlight, resulting in long exposures.

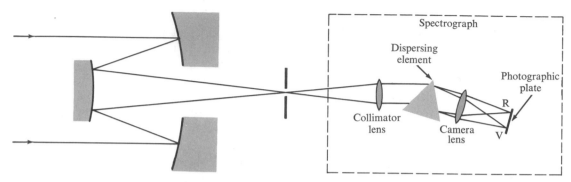

Figure 4–15. *A telescope may analyze the light from a celestial source, when rigged with a spectrograph.*

Nevertheless, though it is expensive in telescope time, spectroscopy is vigorously pursued, because spectrogram yields the greatest possible amount of information about the physical nature of an object.

Since spectroscopy can yield so much information, there are actually various types of spectroscopic investigation which can be conducted. In fact, there are a variety of spectroscopic instruments in use in astronomy. For studies in which the spectrum of an object is needed to determine its temperature, or perhaps a rough index of its size, or to establish whether any gross peculiarities exist in its spectrum, a relatively simple spectrograph may suffice. One such design, which mounts on the telescope at the Cassegrain focus, is shown in Figure 4–16. The most demanding spectroscopic work generally involves spreading out the light in the spectrum to the greatest practical degree. This is known as *high-dispersion* spectroscopy. These spectrographs are generally very large, and cannot be mounted directly at the telescope. Such devices require a focus for a moving celestial object which remains in a fixed position, as the telescope follows the stars. The *Coudé focus* previously described accomplishes this by use of extra mirrors.

## Practical Considerations for Observing

### Star Charts

As we will see below, there are many and varied observations which can be performed, determined by the interests and inclinations of the observer. However, before any observations can be

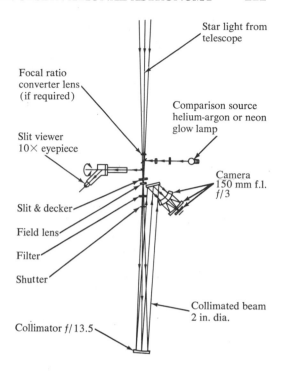

Star light from telescope

Focal ratio converter lens (if required)

Comparison source helium-argon or neon glow lamp

Slit viewer 10× eyepiece

Camera 150 mm f.l. f/3

Slit & decker

Field lens

Filter

Shutter

Collimated beam 2 in. dia.

Collimator f/13.5

Figure 4–16. *Left: a 26-inch (66-cm) tele-scope equipped with an efficient spectro-graph. Right: the optical design incorporates a grating, ruled with many lines, as the dis-persing element.*

made, the objects must first be found at the telescope. This is usually accomplished with a finding chart, made with the aid of complete charts of the sky. The finding chart contains, besides a reproduction of the star-field around the object of interest, positional information in the form of the right ascension and declination of the star, and a number giving the object's apparent brightness (to be discussed later). The position for the object refers to a given date. If it is more than a few years old, it should be corrected for precession—the slow re-orienting of the earth's axis on the sky, which slides the coordinate system with reference to the stars. The stars, of course, are stationary (except for very small motions) but our coordinate system for them is attached to the non-stationary earth's axis, necessitating the change. The telescope has two large mechanical circles, or if recently constructed, electronic readouts giving its pointing direction, so the star-field is first located by setting the telescope to the object's current position. The star-field observed in the telescope's smaller "finder" is then checked against the star chart.

For fairly bright stars, such as those visible in a typical finder of about 6-inch (15-centimeter) diameter, the finding charts may be conveniently taken from a number of popular and readily available charts. Still useful are the hand-constructed sky maps made from visual observations of the whole northern sky in the mid-1800s in Bonn, Germany, known as the "Bonner Durchmusterung," or simply the BD's. Since they were compiled using a small telescope, they are well-suited for use with finder telescopes. Therefore, virtually every star visible through a finder has a BD number, so observers can indicate to each other objects for study, without a need to draw maps. Supplementing the BD charts today are photographic atlases. One useful and low-priced set was made by a German amateur Vehrenberg, photographing with his small telescope. His atlas has become quite popular. For the perusal of the sky, especially galaxies and clusters, the Czech Antonin Bečvar's atlas is very fine, and inexpensive. For stars fainter than can be seen in a small telescope, the best recourse is to the photographic compilation of the sky in the National Georgraphic Society-Palomar Observatory Sky Survey.

### Magnitudes: The Astronomer's Scale of Brightness

When we go out at night to search for a given star or other object, we need to have a measure of the object's *apparent* brightness as well as its position. (Actually, a precise way of specifying this apparent brightness plays an important role in determining the distances to the stars, or galaxies.) The measure of a star's apparent brightness is known as its *apparent magnitude*. Magnitudes have a very old history. Let us recall that the Greek astronomer Hipparchus had divided the stars into six categories of apparent brightness. Such a division was arbitrary and rough, but it served in helping to identify the stars, and gave the astronomers of that time a tool for testing the conventional wisdom that the brightness of the stars never changes. The brightest stars were called stars of the first magnitude; the next brightest were stars of the second magnitude. The sixth magnitude contained those stars just visible to the unaided eye.

The magnitude scale is still used today, but has been refined to great precision. It is now recognized that a magnitude represents a *measure of the ratio* in brightness between a given star and a star, or group of stars, which help define the standard of brightness. The standard is *assigned* the magnitude of 0.0, exactly. Then the magnitude of any other star may be reckoned either from the formula below, or with the help of Figure 4–17, as we will illustrate. The formula from which the figure is taken is:

$$\text{magnitude difference} = -2.5 \times \log \left[ \frac{\text{(brightness of given star)}}{\text{(brightness of the standard)}} \right]$$

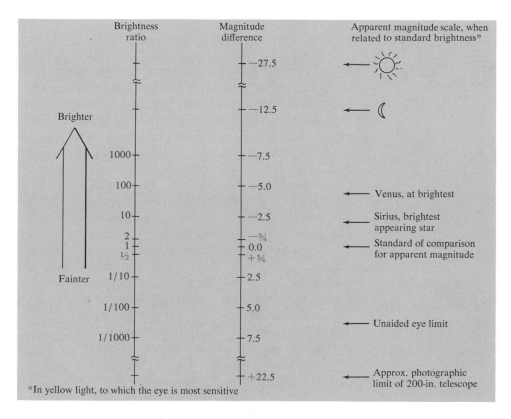

| Brightness ratio | Magnitude difference | Apparent magnitude scale, when related to standard brightness* |
|---|---|---|
| | —27.5 | ← ☀ |
| | —12.5 | ← ☽ |
| 1000 | —7.5 | |
| 100 | —5.0 | ← Venus, at brightest |
| 10 | —2.5 | ← Sirius, brightest appearing star |
| 2 | —¾ | |
| 1 | 0.0 | ← Standard of comparison for apparent magnitude |
| ½ | +¾ | |
| 1/10 | 2.5 | |
| 1/100 | 5.0 | |
| 1/1000 | 7.5 | ← Unaided eye limit |
| | +22.5 | ← Approx. photographic limit of 200-in. telescope |

Brighter

Fainter

*In yellow light, to which the eye is most sensitive

Figure 4–17. *The brightness of two objects expressed as a brightness ratio and as a magnitude difference. When the standard brightness is used as the basis of comparison, the magnitude difference is called an* apparent magnitude.

The negative sign insures that stars fainter than the standard will have positive, increasing magnitudes (as Hipparchus desired), while any object brighter than the standard has a negative magnitude. A star as bright as the standard would be among the brightest stars visible to the unaided eye. So only a few stars and the moon and sun have negative magnitudes. These magnitudes just described are called *apparent magnitudes* because they relate to the brightness of the stars *as we see them,* and are *not* a measure of their *true* brightness. To know the true brightness of a star, we would have to know the distance to it. A few examples will show how the magnitude table illustrated in Figure 4–17 is used.

A star 10 times fainter than the standard brightness has a magnitude of 2.5; a star 100 times, or 10 × 10 times fainter, has a magnitude of 2.5 + 2.5; a star 1000 times, or 10 × 10 × 10 times fainter, has a magnitude of 7.5. In general, we can readily figure the apparent magnitude of a star by breaking any ratio into convenient factors of 10, 2.5, and 2 and by adding each of the magnitudes corresponding to those factors. On occasion, astronomers may be comparing two stars in brightness and refer to their magnitude difference. We can in fact use Figure 4–17 any time we are given the magnitude difference between two objects to figure out their ratio in brightness. Studying the figure, we see that the range of the ratio numbers is quite large, while the range of the magnitude numbers is relatively small. An impersonal instrument, such as an electronic device measuring light intensity directly, would give numbers like the ratios; but the human eye reports sensations of brightness which are compressed like the magnitude numbers. It is by this means that the eye handles the tremendous range of brightness that exists in the everyday world.

Since the stars generally radiate different amounts of energy at different wavelengths, it is necessary, when giving a magnitude, to specify the portion of the spectrum which was used in observing the brightness ratios. There was, of course, no ambiguity in times past, when only the eye was used. Those comparisons were actually made in yellow light; there the eye is most sensitive. Today, photographic plates and light-sensitive electronic devices (photomultipliers) are sensitive to a wide range of wavelengths, so the region of interest can be chosen by placing color filters in the light beam. A standard of brightness, which can be assigned magnitude 0.0 exactly for every wavelength region, needs to be defined. This can form the basis for comparison for all the stars.

It is necessary that investigators agree to use identical groups of color filters and light-sensitive devices when establishing a standard system for brightness measurements in different colors. In this way, measurements of the relative amount of light of each color obtained in observing an object can be compared directly by different observers. The principal system used by astronomers today was established in 1956 by astronomers H. Johnson and W. W. Morgan. They published the results of measurements of many stars, scattered conveniently over the sky, which were intended to serve as standards, or benchmarks for comparison of other stars. The stars were observed in three colors, *ultra-violet*, *blue* and *yellow* (also known as *visual*) light. The measurements of these stars establish, or define, what has become known as the U, B, V system of photometry. Any two measures, in different colors, of the brightness of the same star may be compared in much the same way as one would compare the brightness of two stars observed in the same color. The ratio of brightness may be expressed as a magnitude difference called a *color index*. If we compare ultraviolet to blue light for a star, the result is the magnitude difference for ultraviolet versus blue light, written U-B. Comparison of blue and yellow light leads to a B-V magnitude difference, or color index.

When Johnson and Morgan calculated and plotted U-B and B-V for all the stars they had observed, they found at once that the majority of the stars are confined to a curved line on the plot, called the *two-color diagram.* The most common of the stars lie along a wavy line in the diagram shown in Figure 4–18, if they are in the nearby vicinity of the

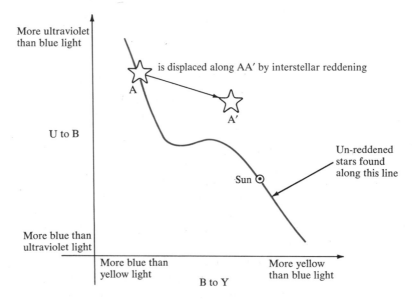

Figure 4–18. *The effect of dust in reddening starlight can be seen when stars are plotted in this diagram. Most unreddened stars lie along the wavy blue line; hotter stars at the top, and cooler stars at the bottom. Reddening displaces stars from the wavy line along lines which are parallel to AA'.*

sun. The hottest stars lie at the top left, the coolest at bottom right in the diagram. Stars observed at a greater distance often do not lie along the line, but are displaced from it along the direction AA'. This happens because starlight, passing the intervening *dust* of interstellar space, has been "reddened" just as the light of the sun becomes reddened as it sets. Thus astronomers can use the measures of the colors of the stars to investigate the intervening space.

Another example of the utility of such a system derives from the fundamental idea that the color of a heated object depends on its temperature. Since stars are heated objects, we expect that, at least roughly speaking, their colors also depend on their temperature. Consequently, the coolest stars have more yellow light relative to blue, and more blue light relative to ultraviolet. The opposite is true of the hottest stars. We shall utilize this information in Chapter 6.

### Timing of Observations

Stars in a limited region near the celestial pole may be studied at any season of the year, but this is not true of those which rise and set. As the earth traverses its orbit, these stars opposite the sun will be the ones visible for the longest period of time during the night. During the course of the year, then, there is an observing season for each portion of the sky in best view; new observable objects shift into prime observing position as others drift out. For example, the stars of the constellation Orion are best studied in the winter.

The phases of the moon also exert considerable influence on the choice of objects to be observed on a given night. A bright moon illuminates the sky, as sunlight does, but on a smaller scale. Away from city lights, it is easy to see that a full moon gives the sky a noticeably bluish cast. Its own illumination is sufficient to allow reading newsprint by dark-adapted persons. Scattered moonlight makes it difficult to observe stars fainter than would be visible through a small telescope. Observatories generally divide their observing time into dark-time and bright-time, about two weeks each. The very brightest nights near full moon must be devoted to bright objects, such as the moon and some of the planets, and are also used for testing various devices and telescope systems.

There is also a class of astronomical phenomena which run on their own time. Eclipses of the sun and moon, and the occultations of stars by the moon and by the planets and their satellites are examples of this class. Those stars which are periodically variable in their light or spectral characteristics are also members. To observe these we must catch them at the right time.

Then, too, there is always the unexpected. Through a specially designed network reports continually reach observatories of various important transient phenomena. These include the discoveries of new comets and supernovas, and dying and violently exploding stars. Thus, scheduling of telescope time is not such a simple matter.

### Physical Conditions of Observing

The physical requirements of an observing session can be very demanding, especially in the winter months. The desired clear nights tend to be the coldest of the season, as there is no cloud blanket helping to prevent the escape of the heat delivered to the earth during the day.

The fresh wind which is also characteristic of those nights adds its effect to the chill. Here, the dome may offer some protection. However, the high altitude of many observatories is not helpful to this situation. The observer must dress very warmly, yet have sufficient freedom of movement to operate the necessary gear. Often gloves become awkward when attending to the constant adjustments of the instrumentation. Inevitably, the eyepiece is breathed on and fogs over. While successful observation demands constant attention, there is little gross body movement, which would aid in keeping warm. Naturally, the question arises, Why not heat the dome? The answer is that the resulting hot air would rise and drift around the telescope, creating the type of images one sees when looking at a scene over a road surface on a hot day. At the expense of dragging around some wire, the problem can be solved using electrically heated flight suits.

Although some telescopes are relatively easy to use, others are not. The smallest are easily managed. The largest, as seen from color plate 8, are also convenient for the observer and have fully automatic motions. However, there are instruments which are large, but are not motorized. They must be moved to different directions on the sky by hand. While the instruments are balanced, even to balancing the instrument load, they still possess a large inertia and can be difficult to shift around quickly. Further, some instruments, when pointing in certain directions, virtually require the observer to stand on his head to reach the eyepiece.

A few telescopes, in fact, are inherently hazardous to use. Consider a 36-inch reflector used at its Newtonian focus. When the instrument is pointing to the zenith, the eyepiece may be thirty feet above the floor of the dome. In fact, astronomers have slipped from the observing platform or ladder using such instruments. One reflector even had a reputation as the "great refractor" since so many bones were broken at it.

With all these discomforts, then, it is impressive to learn that virtually all professional telescopes are solidly booked up and used every clear night. The progress of astronomy is testimony to the fact that a large number of dedicated people found that the reward of securing data outweighed any difficulties involved.


## Limitations to Telescopic Observation

### Atmospheric Limitations

Not all radiation from celestial objects survives passage through the earth's atmosphere. Visible light, for example, typically suffers a 25 to 50 percent decrease in intensity due to absorption and scattering by molecules of the air, as well as dust and dirt. For ultraviolet light, neighboring the visible region of the spectrum, the atmosphere is almost totally opaque. While this screening protects us against the harmful components of sunlight, it prevents ground-based observations of the ultraviolet light from stars and galaxies, an important object to study. For the still-shorter X-rays and gamma rays, we must also get above the atmosphere. The atmosphere is also generally not transparent to infrared radiation, but there are limited regions of high transparency, known as "windows." There are also generous windows for radio waves.

Even if the atmosphere is relatively transparent to visible light, it blurs the telescopic images somewhat. On a hot day, one notices this effect by eye (Figure 4–19). It is caused by the mixing of hot and cold

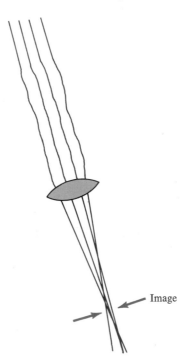

Image

Figure 4–19. *An exaggerated view of the effect of atmospheric distortion on the light from a star; the star image is enlarged in the telescope.*

blobs of air. Light rays traveling through such blobs are bent from their original directions. As Figure 4–19 shows, the rays of starlight are then not parallel when they strike the mirror; thus they will not be focused at one point. The image is instead a blurry disc. Depending on the telescope site, the blur may be from 1/2 arc-sec to 3 arc-secs or perhaps more. This not only puffs up star images, but it tends to wash out the details of planetary surfaces, galaxies, and any other extended or nonstellar object. This problem imposes serious limitations on some important astronomical studies. Telescopes above the atmosphere are not subject to this limitation, which means some of these difficulties may be alleviated (Figure 4–20).

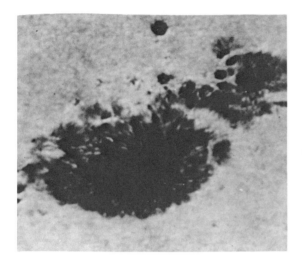 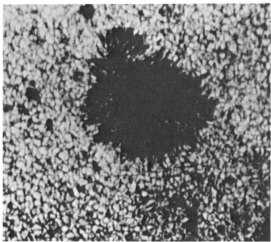

Figure 4–20. *Graphic illustration of the deleterious effects of the atmosphere on telescopic images: the sun (a) as photographed from the ground; (b) from high-altitude balloon.*

## Diffraction

Even when telescopes are carried above the earth's atmosphere, there will still be a limit to the amount of detail present in telescopic images. For example, all stars, except the sun, are so distant that the star images are incredibly small. We cannot see their discs, and unfortunately we see only the effects of diffraction of starlight. Figure 4–21(a) shows an example of this diffraction. Diffraction occurs in telescopes because light waves impinging on an objective encounter its edge, akin to water waves penerating a gap in a sea wall. Parts of the wave, especially near the edge, are bent slightly away from their original direction and cannot be focused at one point by the objective. Because the regions around the edge of an objective are relatively more important in small objectives than in larger ones, we could expect that diffraction would be more severe for smaller telescopes than with larger ones. This is the case. Figure 4–21(b) shows that diffraction can mask the double nature of very close stars, and that a larger telescope can help to alleviate diffrac-

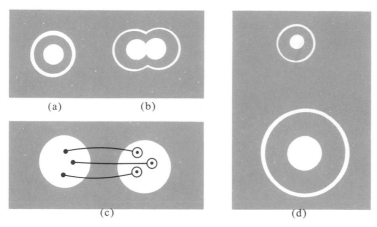

Figure 4–21. *The effects of diffraction at the telescope aperture on stars and extended objects: (a) on the image of a single star, (b) on the image of a close double, and (c) on the image of a planet. Each point of the extended object behaves as the image of a single star, illustrating that diffraction limits the resolution of planetary detail. (d) The size of the diffraction pattern is inversely proportional to the telescope diameter; e.g., the diffraction disc of a 4-inch telescope (top) is ¼ that of a 1-inch telescope (bottom).*

tion's effects. Diffraction affects every type of astronomical observation, not just that of observing close double stars. The ultimate effect of diffraction is loss of ability to resolve detail. For example, suppose we regard a planet, for telescopic purposes, as an object consisting of many very tiny points of light. As Figure 4–21(c) shows, each of these points will not be imaged as a real point, but as a diffraction pattern, so our image will consist of many overlapping diffraction patterns. As a result, details smaller than a certain size will be washed out.

The severity of diffraction actually depends on the wavelength of radiation involved, in addition to the size of the objective. Shorter waves are less severely affected. Actually, the quantity which is crucial in diffraction effects is the *ratio* of the wavelength of radiation to the diameter of the aperture, $D$. The angular size of the "diffraction disc," the central region seen in Figure 4–21(a), is given by

$$\theta \text{ arc-sec} = 2 \times 10^5 \frac{\lambda}{D} \qquad (\lambda, D \text{ both in the } same \text{ units})$$

Let us calculate this angle for a 10-cm (4-inch) telescope and yellow light ($\lambda = 5.5 \times 10^{-5}$ cm):

$$\theta \text{ arc-sec} = 2 \times 10^5 \times 5.5 \times \frac{10^{-5}}{10} = 1.1 \text{ arc-sec}$$

At the moon's distance, 1 arc-sec is about 1 km (1/2 mile).

Diffraction is especially important for radio telescopes, whose dishlike mirrors bring radio waves to a focus. A typical wavelength important to radio astronomers would be 21 cm. Let us calculate $\theta$ for the largest mirror in use, a 1,000-foot dish (1,000 ft. $\times$ 12 inch $\times$ 2.54 cm):

$$\theta \text{ arc-sec} = 2 \times 10^5 \times \frac{21 \text{ cm}}{1,000 \times 12 \times 2.54 \text{ cm}} = 1,300 \text{ arc-sec}$$
$$\text{(about } 1/3°)$$

Thus, a single radio telescope does not see the sky with much resolution.


## Progress in Above-Atmosphere Observation


### Balloons and Rockets

For reconnaisance of selected regions of the sky, balloons and rockets are an economical supplement to satellites. Rocket payloads may be sent higher than balloons but stay aloft for limited amounts of time. Rocket studies of the X-ray sky are presently a very active program. Balloons have been used to lift telescopes above the earth's blurring atmosphere, as noted previously. They also are being currently used in ultraviolet studies of the brighter stars, and for infrared mapping of the sky in wavelengths where the atmosphere has no windows.


### Satellites

The space age has given astronomers a great opportunity for intense studies of the sky in spectral regions previously inaccessible. Three satellites playing important roles to date have been the OAO-2 (Orbiting Astronomical Observatory), launched in 1966, for ultraviolet studies; the X-ray satellite UHURU (Swahili for "freedom"—UHURU was launched from Africa); and the Copernicus ultraviolet satellite. (See Figure 4–22.) A number of smaller satellites especially designed to study the sun, the OSO series (Orbiting Solar Observatory), have also been orbited.

Figure 4–22. *Top left: the OAO-2 rocketed into orbit in 1966 with four telescopes aboard. Its 8-inch instruments examined the ultraviolet light of brighter stars. Top right: the OAO-3, the Copernicus satellite, carried a 36-inch telescope and was able to reach much fainter stars for ultraviolet studies. Right: the UHURU x-ray satellite.*

### OAO–2

The OAO–2 made important contributions to the study of hot stars, which emit most of their energy in the ultraviolet. It was able to perform both photoelectric photometry and spectroscopy, but due to its aperture of only eight inches it was limited to the brighter stars. However, many stars still could be observed, and the distribution of their energy in the ultraviolet was compared to that predicted from models of stellar atmospheres calculated from theory. These checks showed that the theory was on the right track. Single stars with peculiar characteristics and binary stars have also been studied with profit. The star observations could also be utilized to study the absorption of starlight by the interstellar medium. While it was well-known that stars contributed to enriching the interstellar medium, the OAO-2 data showed that certain hot stars sloughed off material at a far greater rate than was estimated from ground-based observations. Since stars are also created from the material of the interstellar medium, this information is of fundamental importance to studies of the evolution of stars. Some of the brighter galaxies were also studied and found to be unexpectedly bright in ultraviolet light. The reason is not yet understood.

*UHURU*

The UHURU satellite made the first reconnaisance of the X-ray sky, with epoch-making results. The number and types of objects detected exceeded even optimistic expectations. A new class of binary stars, the X-ray binaries, was revealed, in which one of the binary members is an X-ray emitter. In several of these, the other star is visible optically. The visible members of the X-ray binaries vary in light, and photometry is yielding interesting results from them. The X-ray companions may be neutron stars, or perhaps, in a few cases, black holes. Both of these are end products of a star's evolution, to be discussed in Chapter 9. UHURU also detected a type of X-ray object which is transient; but these objects have not yet been identified optically, so their nature is still in question. In addition, there is a general background of X-radiation which is related to our own stellar system, the Galaxy. UHURU has also found X-rays coming from positions on the sky coinciding with distant peculiar galaxies and from clusters of galaxies.

From UHURU, astronomers are learning new facts which will help in the understanding of many types of celestial objects. New questions are bound to arise also, and this is what keeps science interesting: nature is always one step ahead.

## Ground-Based Astronomy in the Nonvisible Region

### Infrared Studies of the Sky

If we look at the spectrum of a heated object, we find that in the lower range of the temperatures of the stars [for example, below 3,500°K (about 7,000°F)], the greater portion of the energy is given off in the infrared portion of the spectrum. Now, suppose there exist stars at temperatures of roughly a thousand degrees, or below. These would put out only a minute fraction of energy in the visible region. Such objects might be so faint that they would be difficult to discover by their visible light even in the largest telescopes, but they might be detected by sensing their infrared, or heat, radiation. This rationale prompted astronomers to begin probing the sky with detectors sensitive to infrared rays. Not all of the infrared energy from objects in space can penetrate the atmosphere; but, as we discussed earlier, there are limited regions of the infrared spectrum (known as "windows") where the atmosphere is quite transparent.

Observations in the infrared presents its own unique problems, because it is heat radiation which is sought, whereas all objects emit some heat of their own unless at a temperature of absolute zero. We might imagine an analogy of carrying out visual observations with all of our equipment glowing with light, including the telescope, the dome, and even the instruments attached to the telescope. Therefore, special techniques are used with an infrared detector to screen out unwanted heat from its surroundings so that the unwanted heat does not completely overwhelm the feeble amount of heat collected by the telescope from the celestial source. One unique advantage of infrared astronomy, however, is the ability to make useful observations in broad daylight. This is possible because daytime is considerably darker in the infrared than in visible light. To understand this, consider why the sky is blue. The blue light coming from a patch of sky is derived from sunlight, which has been scattered in our direction by air molecules. It is well understood that the air scatters short wavelengths of light better than long wavelengths; hence, the light reaching the eye from a patch of sky is rich in blue light, as shown in Figure 4–23. Also consider that when the sun sets, it appears red because more blue rays are scattered from the sunlight than red rays. The theory of scattering tells us that the efficiency of air molecules to scatter light is proportional to $(1/\text{wavelength})^4$. If we consider yellow light, of wavelength 5,000 Å, and near-infrared light of 10,000 Å, the latter will be scattered only $(1/2)^4 = 1/16$ of the visible light. The less the scattering, the less bright will be a patch of sky. Since infrared astronomy is routinely conducted at wavelengths of about 100,000 Å, the sky is as dark in the daytime as at night. Thus we have the unusual spectacle of the far-infrared observers busily gathering telescope data in the middle of the day.

*Some Achievements of Infrared Astronomy*

As noted above, an important rationale for the development of infrared astronomy was to search for cool objects not detectable in the visual region. Did the new work meet this expectation? It did indeed, and more. In fact, every time a new region of the spectrum has been opened up for investigation, an unexpectedly rich harvest is reaped, full of surprises. The new discoveries in infrared astronomy do include cool objects. The most intriguing of these was found in the Orion Nebula. What is exciting about this discovery is that theoreticians have predicted that a star in its very earliest phases would appear as a very cool and large body, contracting down on its way to "stardom." The Orion Nebula does have other visible stars in it, and astronomers already knew these must be extremely young. (The determination of the ages

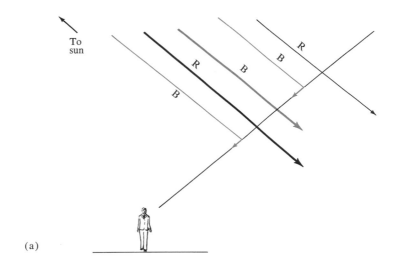

(a)

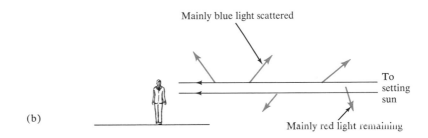

(b)

Figure 4–23. *(a) Sunlight scattering into an observer's line of sight. Much more of the blue light than the red is scattered, making the sky blue. (b) Blue light is severely scattered at sunrises and sunsets due to the light's long path through the denser lower atmosphere.*

of stars will be discussed later.) It was therefore extremely rewarding to be able to find the progenitors of visible stars by detecting their infrared radiation. This, of course, helps to confirm that theoreticians are on the right track in the research on star formation.

A great surprise, however, was the finding that some galaxies, those largest aggregations of stars, were also profuse emitters of infrared radiation. Even though such studies are only in their infancy, they have already produced evidence that galaxies can produce far more infrared energy than visible energy; but the mechanism for this process of energy production is not at all understood.

Infrared studies have also been of great value in studies of the planets and satellites of the solar system, especially in the determination of their temperatures. One planetary temperature measurement has led to a great deal of theorizing and speculation. This is the unusually high temperature indicated for the planet Jupiter, which we shall discuss in the next chapter. These are some of the very impressive contributions of infrared astronomy—they indicate a bright future for this young science.

## Radio Astronomy

In 1931, the experiments of Karl Jansky of the Bell Telephone Laboratories ushered in the field known as radio astronomy. While it might have been expected merely to complement somewhat the work of classical, or optical, astronomy, it has instead brought to astronomy its own unique dimension of information. Radio astronomy explores phenomena not available to the optical astronomer, and it touches on different phases of those things which the optical worker also studies. The "eye" of the radio telescope, working at wavelengths longer than about a millimeter, sees an intriguingly different picture of the sky than its optical counterpart.

Radio astronomy has made revolutionary contributions to the fields of solar system studies, the interstellar medium, galaxies, and cosmology. Furthermore, it is only in radio studies of the solar system bodies that we find exception to the statement that observational astronomy is a passive science. Objects may be probed by transmitting powerful bursts of radiation from the telescope, and dissecting the returning echoes.

### The Radio Telescope

The workings of a radio telescope, shown in Figure 4–24, strongly suggest its similarity to an optical telescope. There are, in fact, even Cassegrain radio telescopes, as shown. There is, however, an important difference between radio and optical instruments connected with the phenomenon of diffraction, which is more important for the radio telescope, as explained earlier. Because an individual radio telescope has such poor ability to resolve detail, it would be better to regard it as a collector of energy (the amount of which is of interest) rather than as an image-former. The great size of a radio dish is an attempt, not only to beat diffraction, but also to collect as large an amount of energy as possible. The largest radio telescope in the world is located at the Arecibo Ionospheric Observatory in Puerto Rico, which features a fixed dish of 1,000

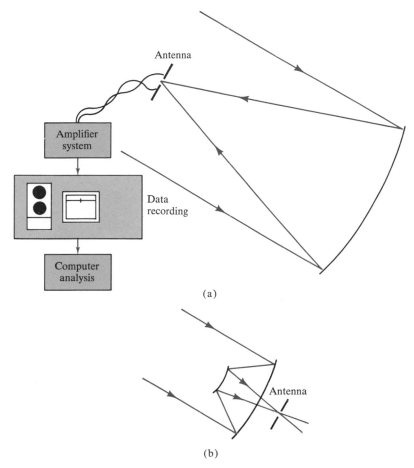

(a)

(b)

Figure 4–24. (a) The elements of a radio telescope. The signal from the antenna is amplified, and then recorded (for example, on tape or on a chart recorder). Data may be further subject to computer analysis. (b) Cassegrain configuration for a radio telescope.

feet diameter. For short intervals of time, it is able to track celestial objects by moving its antenna, which is at the focus of the metal mirror, some 400 feet above the ground. Construction costs were kept to a minimum by cleverly locating the dish in a natural depression among three mountains (Figure 4–25). The largest fully steerable dish is a 100-meter (300-foot) West German telescope near Bonn, operated by the Federal Republic. A telescope of equal size but steerable only in the north-south direction is the 300-foot telescope at Greenbank, West Virginia, shown in Figure 4–26. Other large fully steerable dishes are

Figure 4–25. *The Arecibo 1000-foot (300-meter) radio telescope, located in an artificially enlarged depression in the Puerto Rican hills. It is the world's largest radio telescope and a pioneer in radar studies of the solar system.*

located in England and Australia, at Jodrell Bank and Parkes, respectively. Scattered about the world are numerous instruments of the 100-foot class and smaller. All of these instruments can operate around the clock and in most kinds of weather.

We have discussed previously that the resolution of a single radio telescope falls far short of that provided by even a small optical instrument. However, two or more radio telescopes separated by larger distances may be operated in a way that produces a resolution that no optical instrument can hope to match (at least for the present). Cur-

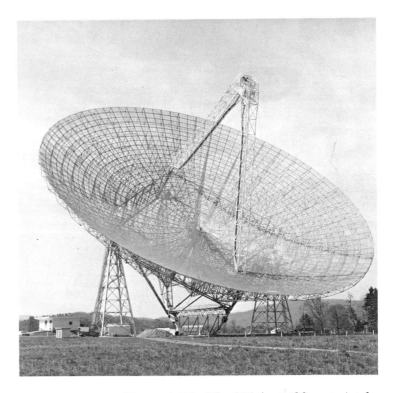

Figure 4–26. *The 300-foot (90-meter) telescope of the National Radio Astronomy Observatory at Greenbank, West Virginia. Instead of tracking particular celestial objects, the instrument is set in the north-south direction and the rotating earth carries the beam eastward around the sky.*

rently, the record for resolution is 0.0005 arc-sec (with a possibility for improvement). The secret of this precise resolution lies in the ability of modern electronics to follow the exact undulations of the radio waves as they are received, even though the frequency of the waves may be some hundreds of billions of time per second. Of course, to do this, one needs the best clock available—an "atomic clock." It keeps time to a few thousandths of a second a year, while subdividing the second into hundreds of billions of parts. The radio signals at the two telescopes and the time of observation are recorded together on magnetic tape, and the tape records may be compared, moment for moment, in a computer.

While detailed descriptions of the process would be beyond the scope of this book, we can describe fairly simply what effect is being accomplished. We have mathematically, and for practical purposes, created a radio telescope with a resolution which depends on the *separation* of the two telescopes. By comparison, the resolution of a single telescope depends on the separation of the two most distant points of the dish, or mirror, that is, its diameter. The spectacular resolution achieved at the present is due to the use of intercontinental separations of the two instruments. Remember that the resolution depends on the ratio of this distance to the wavelength of the radiation concerned. Figure 4–27 shows a map of a small region of the sky obtained in this way. (Note the scale.)

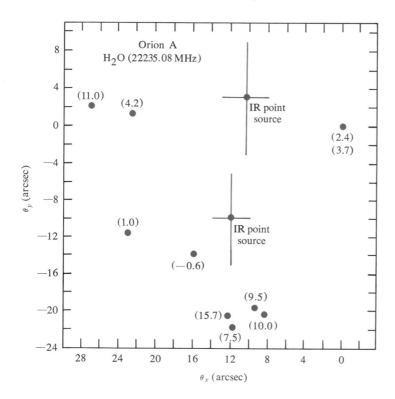

Figure 4–27. *A small region of the sky, illustrating the high-resolution capabilities of an interferometer with intercontinental baseline.*

Instead of a typical intercontinental distance of 6,000 miles, consider what resolution would result from having one of the telescopes on the moon. Two telescopes operated in this way are known as an *interferometer*. Two observations about the interferometer are worth noting. First, the resolution is only enhanced in one direction: if the two instruments are located E-W, then resolution on the sky in the E-W direction improves. On the other hand, the N-S resolution is unchanged from that of a single instrument. Adding a third instrument to the interferometer in the south direction, for example, could be used to improve the N-S resolution. Again, although the interferometer behaves like a very large radio telescope, its collecting area is merely that of two or three telescopes. Now, if we were to string out a number of 100-foot radio telescopes over an area of several miles, we would have an instrument with quite a large collecting area, and with a fairly good resolution, equivalent to one telescope several miles in size. In effect, we are creating a large dish with holes in it, but it is still a large dish.

This type of telescope grouping has been built at Westerbork, in the Netherlands. It will be used especially for studying hydrogen gas distribution of the Milky Way Galaxy, and for studies of external galaxies. The Westerbork telescope produces pictures of the sky, computer-made from the collected data. A smaller instrument, designed especially to study the sun, is located at Culgoora, in Australia. Images of the sun, in radio waves, are produced on a television screen by the system computer. In the United States, a very large array (known imaginatively as the VLA) is under construction. It will have 27 telescopes 82 feet in size which can be moved to any of 100 observing stations along a 37-mile double track railway laid out like a Y (Figure 4–28).

Figure 4–28. *Artist's conception of the Very Large Array (VLA) radio interferometer, to be located in New Mexico near the continental divide. The completion date is 1981.*

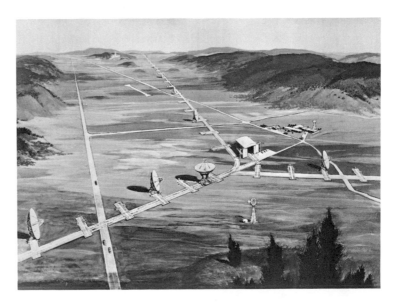

*Radar Astronomy*

In solar system studies, radio astronomy is in the unique and enviable position of being able to take an active role. This involves the use of radar—the detection and analysis of echoes of the signal sent by the transmitter. The principle has, of course, been familiar for many years, but as applied in astronomical studies, it is different in many ways. First, the level of the returned signals is considerably weaker than in typical earth-based radar. When a signal is sent to a planet or satellite, only an incredibly small fraction of that power is returned to the sender. We get some idea of the reason for this from Figure 4–29. The waves, of necessity, spread out when they leave the telescope; thus only a small fraction of the waves will be reflected by the planet. Then, only a small fraction of that will be intercepted by the dish, as the wave, spreading out on the return journey, reaches the earth. The tiny signal must be found among a cacophony of cosmic and man-made static and static (or *noise*) in the receiver.

The situation is much like trying to carry out a long-distance conversation with a whispering friend who is at a party being held at an intersection and construction site. An individual sentence uttered by the friend might be inaudible due to the noise. Suppose, however, that the friend were to repeat the same sentence over and over, while we recorded it on tape in such a way as to overlay it on the same place on the tape. In this manner, the sentence could be made to stand out from the background noise. In the same way, the radio astronomer beams many pulses of radio energy to the object of interest, adding the echoes in the manner described. What does the echo tell us about the object?

In Figure 4–29 we see that the signal impinging on the celestial object strikes the part nearest earth first, then successively larger rings surrounding the initial point reflect the wave. We can thus understand how the length of the echo can tell us the size of the body. Roughly speaking, the time between the beginning and the end of the return signal is the time elapsed between the wave's first striking the body and finally reflecting from the limb or edge. The distance traveled is the object's radius, and the speed is the speed of light. Therefore, the formula for the radius should be $R = c \times$ *the time interval*. From the diagram of Figure 4–30 it is easy to see how we can find the rotational speed of the body. For simplicity we consider the planet not to be moving along the line of sight, but only rotating about an axis perpendicular to the radio wave's direction. The signal which leaves the transmitter is a pure tone, and may be considered a single frequency. Consider the last reflections from the planet or satellite, which come from its limb region. Because of rotation, one limb approaches the observer and the other recedes. The

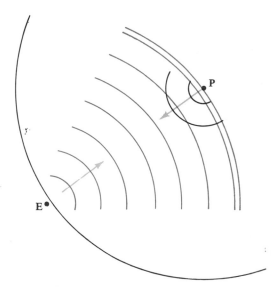

Figure 4–29. *Only a minuscule fraction of the power beamed to a planet (P) from earth (E) is returned, due to the spreading of the beam.*

Doppler effect will therefore cause the trailing ends of the return signal to be split in frequency, and from this split we can deduce the rotational speed of the celestial body. Some of the remarkable results from the application of this technique will be discussed in Chapter 5. The techniques of echo analysis have been subject to the most complete refinement; at the present time it is possible to produce a map of a planetary surface with quite a surprising amount of detail, as shown in Figure 4–31.

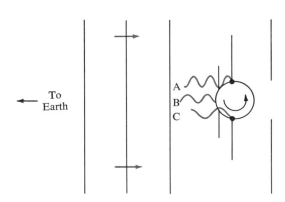

Figure 4–30. *Radio waves reflected from the limbs of a planet are Doppler-shifted. The wavelength at A is shorter than that at B; at C, the wavelength is longer.*

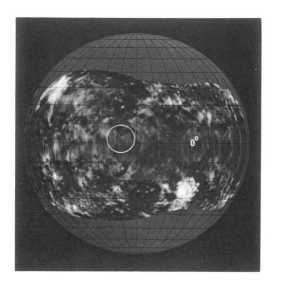

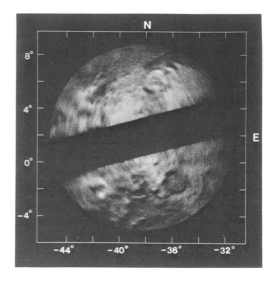

Figure 4–31. *The planetary surface of Venus revealed from its radio reflections. The heavily cratered surface (left) is surprising in view of the thick atmosphere, which should induce a great deal of erosion. An enlarged view of the circled region is shown at right.*

Questions

1.  What are the ingredients for a good telescope site?

2.  What are some of the different modes in which a telescope can be used?

3.  Suppose a 15-inch (38-centimeter) telescope is limited to working on stars brighter than a given magnitude. What then might be the limit for a telescope of 30 inches (74 cm)? (Hint: A 30-inch gathers 4 times more light.)

4.  If the difference in magnitude between two objects is 2.5, what is the ratio of their brightness? If two objects have a ratio of two-to-one in brightness, what is their difference in magnitude?

5.  What are the limitations on telescopic observations from the ground? From space?

6.  In what sense would one regard a radio telescope as different from an optical one?

7.  What is an interferometer? Why does it have such good resolution compared to a single instrument?

8.  Test the theory of diffraction. Draw a map of the moon from your own observations (the moon is 1,800 arc-sec across) and estimate the angular size of the smallest details you have seen. If the eye's aperture is about 7 mm, or 0.7 cm, what should be its limiting angular resolution? How does this compare to your observing results?

# PART

# 2

# PLANETS AND STARS

**5**  The Solar System

**6**  Basic and Observable Properties of Stars

**7**  Determining the Distances of Stars

**8**  Finding the Basic Properties of Stars

**9**  Stellar Evolution

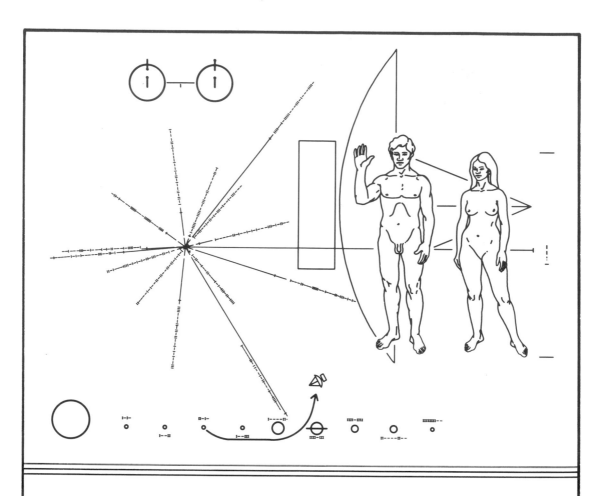

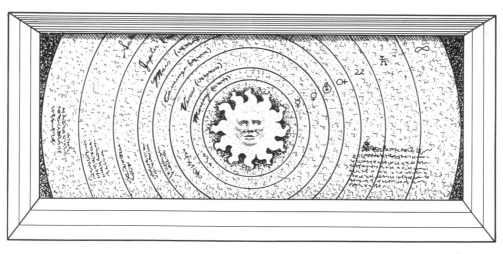

# 5 The Solar System

## Introduction

The solar system consists of the Sun and all the objects in orbit about it. It is our immediate astronomical neighborhood. Our state of knowledge regarding these bodies could be expected to have reached a greater level of sophistication than that of stellar or galactic astronomy merely because of their proximity. Yet this is not so. The glowing gases of the stellar surface practically advertise their temperature. With some effort, much more can be learned about the pressure and composition of these outer stellar layers. Furthermore, we know more about the interiors of most stars than we know of the earth, because the gaseous matter of the stars display a simple relation between temperature, pressure, and density. Solid matter, the stuff of which the planets are made, requires a much more complicated relation, which is not completely understood at this time. The theories of the internal structure of the planets do not get much help from experimenters. Testing solids at pressures corresponding to planetary interiors is not yet a laboratory possibility.

However, planetary astronomy has quickly matured due to the technological progress of the last decade. With large and sophisticated radio telescopes, infrared devices, and bold space probes, the investigations are carried directly to the vicinity of the planetary bodies.

Much effort is being put into the study of the solar system's origin. This will tell us how our planetary system came to be. A big question is whether or not these results could make mandatory the conclusion that we are not alone.

### Overall Structure of the Solar System

When we look at the distribution of the motions of all the objects orbiting the Sun, we discern two quite different groups. The nine major planets, and myriad minor ones, all travel about the Sun in roughly circular orbits. The orbits of these planets lie, in general, close to the plane of Earth's revolution about the Sun. Thus this system is disc-shaped, and we may refer to it as the *flat system.* In contrast to this rather orderly arrangement, the comets have orbits which are generally very long ellipses, often larger than the orbits of the outermost planets, but which can approach the Sun closer than Mercury's orbit. The orbits of the comets are not confined to the plane of the ecliptic and may make any angle with it. Thus, we often see comets which seem to be in retrograde motion about the Sun. We may say that the comets form a *spherical system,* by contrast with the flat system.

The space between the planets is filled with gas and some dust. This is the interplanetary medium, material boiled off the Sun's surface. The motion of this gas, which travels generally away from the Sun, is governed by magnetic fields stretching from the Sun well past the orbit of Earth. We now proceed from this overview to investigate the system of major planets.

### Two Planet Groups

In Table 5–1 we find a very obvious and significant way of dividing the planets into two groups. These groups differ markedly in density. The inner planets all have high densities resembling the earth and may be

## TABLE 5–1

*Planetary Properties*

| Planet and symbol | ORBITAL DATA | | | |
| | Distance from sun (a.u.) | Period of revolution (years) | Inclination to ecliptic | Eccentricity |
| --- | --- | --- | --- | --- |
| Mercury  ☿ | 0.387 | 0.241 | 7° 00′ | 0.206 |
| Venus  ♀ | 0.723 | 0.615 | 3° 24′ | 0.007 |
| Earth  ⊕ | 1.00 | 1.00 | 0° 00′ | 0.017 |
| Mars  ♂ | 1.524 | 1.881 | 1° 51′ | 0.093 |
| Jupiter  ♃ | 5.203 | 11.86 | 1° 19′ | 0.048 |
| Saturn  ♄ | 9.539 | 29.46 | 2° 30′ | 0.056 |
| Uranus  ♅ | 19.18 | 84.01 | 0° 46′ | 0.047 |
| Neptune  ♆ | 30.06 | 164.8 | 1° 46′ | 0.008 |
| Pluto  ♇ | 39.44 | 247.7 | 17° 12′ | 0.250 |

**TABLE 5–1**—*Continued*

## PHYSICAL DATA

| Planet | Mass (earth-masses) | Diameter (earth-diameters, miles, km) | Average density (g/cm³) |
|---|---|---|---|
| Mercury | 0.056 | 0.38; 3015 (4868) | 5.1 |
| Venus | 0.82 | 0.95; 7526 (12,112) | 5.3 |
| Earth | 1.00 | 1.00; 7920 (12,742) | 5.52 |
| Mars | 0.108 | 0.53; 4216 (6800) | 3.94 |
| Jupiter | 318. | 11.20; 88,700 (143,000) | 1.34 |
| Saturn | 95.2 | 9.47; 75,000 (121,000) | 0.70 |
| Uranus | 14.6 | 3.69; 29,000 (47,000) | 1.55 |
| Neptune | 17.3 | 3.50; 28,900 (45,000) | 2.27 |
| Pluto | <.06 | <.5 ; <3700 (<6000) | — |

## TELESCOPIC DATA

| Planet | Period of rotation | Polar flattening (%) | Number of satellites |
|---|---|---|---|
| Mercury | 59$^d$ | 0 | 0 |
| Venus | 243$^d$ | 0 | 0 |
| Earth | 23$^h$ 56$^m$ 04$^s$ | .33 | 1 |
| Mars | 24$^h$ 37$^m$ 23$^s$ | .52 | 2 |
| Jupiter | ~9$^h$ 50$^m$ (var.) | 6.66 | 13 |
| Saturn | ~10$^h$ 25$^m$ (var.) | 10.53 | 10 |
| Uranus | 10$^h$ 45$^m$ | 7.14 | 5 |
| Neptune | 16$^h$ (?) | 2.5 | 2 |
| Pluto | 6.4$^d$ | 0 | 0 |

characteristically referred to as the "terrestrial" planets. The next four planets have low densities. This group is likewise named after its largest member, as the "Jovian" (adjectival form of Jupiter) planets.

This density difference signifies a marked difference in structure, especially in *composition,* between the two groups and therefore must tell us something important about the early phases of formation of the planets. We will return to this point later. It may also be noted that the Jovian group of planets seems to have a monopoly on the important satellite systems.

Let us examine the planets individually. We will see they are each unique.

## The Earth and Moon

### The Earth

Our interest in the earth in this chapter is largely confined to its overall geological properties and related attributes which can be compared to other solar system bodies. In this way, we can get an idea of the value of the results obtained by interplanetary space probes and by direct lunar exploration.

The interior structure of the earth is now well-known from seismic studies. These show that the earth has an outer layer, or *crust,* which is fairly thin, averaging some fifteen miles in thickness. The next region, the *mantle,* is the most extensive, and reaches to a depth of about 3,000 miles, where it meets the earth's *core.* It is known that seismic waves traveling through the interior of the earth do not all penetrate the core. This implies that the core is liquid. There is also some indication of a solid core at the very center. The temperature of the earth increases inward, such that the core is at a temperature of thousands of degrees. The density of the earth also increases inward from a surface value of about 3 g/cm³ (three times that of water).

In order for the overall density of the earth to reach the observed value of 5 g/cm³, it has been shown that the density of the smaller-volume interior must be near 8 g/cm³. It is generally believed, therefore, that the core must consist principally of heavier elements, especially iron. The process by which the heavier elements have sunk to the center is thought to be the heating produced by the decay of radioactive elements during the earth's early history. It is considered that the rotation of the earth stirs the hot metal core and this creates the earth's magnetic field.

The interior heat of the earth is quite considerable and exerts a profound influence on the thin crust, which flows under the tremendous heat and pressure. The continents are like plates floating on the under-layers. This results in the phenomenon known as *continental drift.* At present, for example, Europe and North America are increasing their distance by several inches per year. Another evidence of interior activity is volcanism. However, it is not quite clear yet whether volcanism is an index of the plate-moving activity of a planet.

The questions that the astro-geologists are now asking concern the interior structure of planets: How thick is the crust? Does continental drift occur? Is there a metallic liquid core?

### The Moon

The moon has been a source of wonder and curiosity ever since ancient times. Cults have been formed in its honor; for instance, it is thought that in the British Isles moon worship was stimulated by the fact that, in the far north of Great Britain, the moon could be seen as a circumpolar (never-setting) object. In modern times, the moon's influence is not negligible; being the closest celestial object, it virtually demanded investigation. (Perhaps it would be interesting to speculate on the psychological influence of that body on the growth of astronomy.) Before the space age, a great deal of effort had been expended over the years in getting data on the nature of the moon. This formed a basis from which intelligent planning could be carried out for the manned-exploration phase.

*Earth-Based Study*

Early earth-based studies of the moon furnished details of the shape of the moon's orbit. These studies led finally to the construction of an elaborate theory for the motion of the moon, which responds to gravitational influences of the sun and the planets (mainly Jupiter) as well as the earth. The theory gave the position of the moon against the background stars for any time. Until recently, the position of the moon was, in fact, used to tell the time. Now the atomic clock provides a more uniform and practical standard. Early studies also determined the moon's size and revealed that the side which perpetually faces the earth was bulged slightly (about 1 mile, or 1.5 kilometers) in the earth's direction.

The surface was thoroughly studied telescopically, and maps were constructed. Surface features readily visible from earth include craters, ranges of mountains and the large dark areas called "seas," or *maria*. These are not seas at all, but lava flows. In addition, some craters were noted to have an associated system of divergent rays. Features few in number, but virtually peculiar to the moon, are *rills,* the shallow, long clefts in its surface. (See Figure 5–1.)

The atmosphere of the moon is extremely tenuous. One way to determine this is by measuring the brightness of stars as they drift behind the moon's limb (edge). These observations indicate no gradual absorption, as would be caused by an atmosphere; instead, there is a virtually instantaneous cut-off in light. Very likely, the little atmosphere the moon does have is variable, and is due to spasmodic venting of volcanoes. In 1958, the Russian astronomer Kozyrev detected gases vented from the crater Alphonsus.

The moon may well have had a denser atmosphere earlier in its history. Its current absence of atmosphere is in agreement with calculations showing that its surface gravity of 1/6 the amount of the earth's is not sufficient to retain atmospheric molecules. Without the insulating properties that an atmosphere can provide, the temperatures over the month-long lunar day range from 100°K to 400°K (roughly −230°F to 200°F).

Figure 5–1. *(a) The western hemisphere of the moon, with the Alpine Valley identified (shown enlarged in Figure 5–2).*

Figure 5–1. *(b) The eastern hemisphere of the moon with the rayed crater Tycho identified (shown enlarged in Figure 5–2).*

The moon is important in studies of the early history of the solar system because of its lack of atmosphere. The lack of atmosphere means processes of erosion are vastly slower on the moon than on earth; thus the moon has its ancient conditions still written clearly on its surface. Its craters, due principally to the impact of bodies on its surface, are a very readable record of the past. By contrast, old craters, called *astroblemes,* found on earth are mere shadows of their former selves.

## Spacecraft Studies

*Pre-Apollo.*    Early studies of the moon, preliminary to the Apollo program, include the Ranger, Orbiter, and Surveyor missions. The Ranger probes provided the first spectacular close-up glimpses of the moon's surface as the crafts crash landed. The Orbiters were much more sophisticated and had as their mission a rather complete photographic mapping of the moon. The Orbiter's reconaissance of the moon's far side showed it to be considerably different in appearance from the visible side— it has no maria. The reason for this absence is not known, but very likely the reason is important for understanding the history of the earth-moon system. Figure 5–1(c) shows a typical view of the moon's far side. See also Figure 5–2.

During the Orbiter's circling of the moon, small but unexpected accelerations were noted in the Orbiter's motion. These accelerations were always observed when the craft passed over the maria. From this, it has been concluded that dense clumps of material lay under the maria. These mass concentrations became known as the *mascons* (Figure 5–3). It has been suggested that the mascons are the remnants of large bodies (50–100 km in size) which struck the moon and triggered lava flows. On the other hand, a recent study has determined that the shape of the mascons resembles the maria themselves. The Orbiter's mapping has also been used in crater-count studies to which we will refer later.

The Surveyors were to land at proposed Apollo sites and determine the physical characteristics of the surface, that is, to be sure the surface was free of rubble and could bear the weight of a large spacecraft. We must note here that according to a theory based on radar measures of the lunar surface, the moon was covered everywhere by a 17-meter (50-foot) layer of dust! Actually, a consistency more like wet sand was found. Later Surveyors also tested the composition of the surface and were able to confirm a similarity to the mantle of the earth.

*Apollo.*    On July 20, 1969, the first Apollo mission's lunar exploration began amid considerable excitement. Manned landings of the Apollo scries have gathered selected samples of the moon's surface from a variety of geologic terrain, including the maria and the mountainous "highlands." In addition, the astronauts deployed scientific packages

Figure 5–1. *(c) A portion of the moon's far side, as viewed from the Lunar Orbiter V. The large crater is Moscoviense. The complete lack of "seas" on the moon's far side is noteworthy.*

Figure 5–2. *Lunar Orbiter close-ups of lunar features: the Alpine Valley, situated between Mare Frigoris and Mare Imbrium. The mean lunar distance across the photograph is approximately 286 miles (460 kilometers). The spacecraft altitude was 154 miles (246 kilometers). Right, the rayed crater Tycho (long streak visible in earth-based moon photo preceding), believed to have been created by a giant meteor impact. Glassy materials called* tektites, *found at only a few locations on the earth, are believed to have resulted from this impact.*

cluding seismometers, magnetometers (magnetic field measuring devices), special light reflectors for laser-ranging experiments, and umbrellalike collectors to sample the solar wind. They also took core samples by drilling a meter or so below the surface. The principal idea guiding the selection of landing sites was that the maria, which are relatively free from craters, are younger features on the moon's surface. Presumably, they had not been struck for long periods of time by interplanetary debris and small asteroids. This presumption proved to be correct.

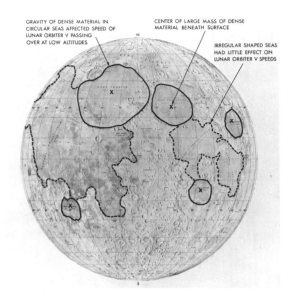

Figure 5–3. *Lunar map showing the locations of the important mascons. They were located by plotting spacecraft acceleration data from the Lunar Orbiter V. There is a coincidence with a lunar sea in all cases.*

*Apollo Results.* The magnetometer measurements from different Apollo flights, if interpreted as measuring a global lunar magnetic field, at first seemed to be in disagreement. However, this was soon explained by the fact that the moon has a very spotty magnetic field (unlike the earth's globally-ordered field, which resembles a bar magnet's field). The fields on the moon seem closely related to distinct surface features, namely craters. The suspicion is that this is due to embedded material, mainly iron, which is especially prominent in *interplanetary* debris recovered on earth (meteorites). Lack of an overall magnetic field very likely signifies the lack of a hot, liquid lunar core.

The laser *retro-reflectors* on the moon have now been used successfully to return laser beams to the earth to determine the moon's distance to well within a meter (or yard). This knowledge will permit refinement of the theory of the moon's orbit. The small, side-to-side oscillations of the moon, known as *librations,* which may now be measured very accurately, will also be studied for clues to the *density structure* of its interior. On earth, lunar laser-ranging from different continents may enable direct detection of continental drift.

Lunar seismic studies show a much lower level of activity for the moon compared to the earth. Unlike the earth, the majority of "moonquakes" are not confined principally to the outer crust. This probably indicates that there is no plate structure, as on earth, and that stiff rock (the lithosphere) extends deep into the moon's interior, perhaps to distances of between 300 km and 1000 km.

Studies of the lunar samples confirm that, in general, the moon's composition resembles the earth's, but with important differences. The moon is lacking in volatile, or easily vaporized, materials and is overabundant in several metals, notably titanium. Many tiny spherules of glass are also contained in the "fines" or dust. This is indicative of a great deal of punishment—melting and pulverizing due to high-speed impacts of particles of all sizes slamming into the surface. There has been considerable mixing due to this process and many Apollo photographs graphically illustrate the large scale results: erosion. (See color plate 4.) Current study shows that the moon's surface erosion is primarily due to an incessant rainfall of sub-millimeter-sized particles, the *micrometeorites*.

## Earth's Companions

### Mercury

The innermost planet of the solar system has been known since ancient times. It was often thought to be two different objects, as some civilizations were not unaware that the fleeting object that set just after the Sun was the same which months later would precede the Sun's rising. Mercury is a very difficult telescopic object, since it must be observed at twilight low to the horizon. The beam of the telescope is then carried long and low over the ground, resulting in poor observation. Mercury never comes closer than 0.6 astronomical units (56,000,000 miles) to Earth, while it exceeds the Moon's linear diameter by only 40 percent. Consequently, it never presents a disc larger than about 6 arcsec. It is thus extremely difficult to discern many surface features on Mercury.

Nevertheless, telescopic maps of the surface have been made. To know the region being observed, it is necessary to know the period of rotation. Otherwise, it would be impossible to fit the individual pictures together. The period of rotation had been assumed to be 88 days, so that Mercury would always keep one face toward the Sun (like the Moon, relative to Earth). Optical work strongly indicated that Mercury had no atmosphere. The most obvious clue was that the planet is considerably fainter than it would be if it reflected like a body with an atmosphere. Only about 7 percent of the sunlight is reflected. (In astronomical terminology, Mercury has an *albedo* of 0.07.) Since the Moon's albedo is also 0.07, Mercury was assumed to be like the Moon.

Strong evidence for a similarity between Mercury and the Moon had also come from radar observations of Mercury made with the Arecibo telescope. The back-scattered radio waves suggested a distribution of slopes reminiscent of the Moon. This study also determined the diameter of Mercury to be $4,864 \pm 4$ kilometers, or $3022 \pm 1$ miles. A most remarkable discovery concerning Mercury was that the rotation period for the planet was found to be 58.7 days. This number is exactly 2/3 the period of revolution, suggesting that the planet's rotation "knows" what the revolution period is. How? The only explanation is that the tidal bulge of the planet is locked-in to its rather elliptical orbit about the sun. The effect is shown in Figure 5–4. Every passage close to the Sun (perihelion) finds the planet facing in the opposite direction. Incidentally, the old visual observations (based on the 88-day period of rotation) were reassembled using the new period. The operation did produce a sensible but ill-defined map. However, we now have spectacular first-hand information about Mercury's surface. Mariner 10 transmitted many pictures of Mercury, including those in Figure 5–5. The strong similarity to the Moon is obvious. The direct photography strongly supports the radio data and augers well for radar studies of perpetually cloud-covered Venus.

While the cratered appearance of the planet was expected, a startling revelation from Mariner was the existence of a very tenuous atmosphere and a weak magnetic field. It had been thought that the high surface temperature and low gravity of Mercury would have allowed its atmosphere to boil away a long time ago. On the other hand, it is conceivable that the planet is still venting off its interior gases, thus continually accounting for any loss. The existence of a magnetic field was a surprise because magnetism in a planet is thought to be connected with the existence of a molten planetary core of iron, which is stirred by the axial rotation of the body. Mercury's slow rotation was previously thought to be too inefficient to stir any core effectively, but this assumption is evidently wrong. Mercury's magnetic field is like that of the earth's. Earth's magnetic field is shaped as if a large bar magnet were lying within the planet, parallel to its rotational axis. Unlike Earth, however, whose magnetic axis passes close to the rotational axis, the magnetic axis of Mercury lies at a distance of 1/2 radius from the center.

Consideration of all the data gathered by Mariner 10 suggests the following capsule history for Mercury. Soon after its birth, the planet suffered intense bombardment by interplanetary debris, which was very

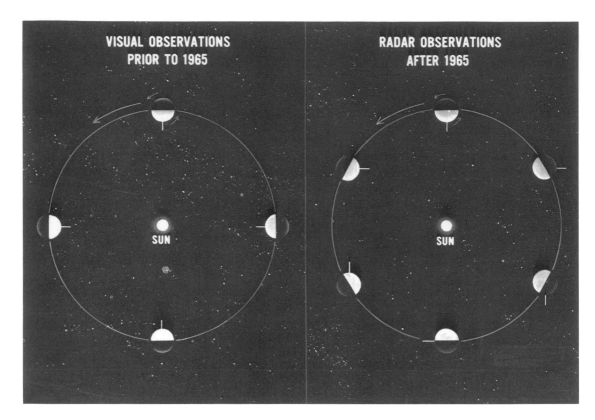

Figure 5–4. *Radar observations determined that the rotation period of Mercury is two-thirds its period of revolution. The planet's orbit, represented here by a circle, is in fact noticeably elliptical.*

much more extensive than at present. This bombardment heated the body, which melted throughout, so that the denser materials (mostly iron) sunk to form a core. Such a change is called *differentiation*. (For example, the earth is a differentiated body, but the moon is not.) In differentiating, the volume of the center part would have shrunk, causing the skin to undergo a shrinking (the dried fruit effect). Mercury's scarps, or ridges about 3 kilometers (about 1.5 miles) high, extending in some cases 500 kilometers (about 350 miles), may have been formed in this way.

Quite likely, Mercury's history may not be done full justice by this simple picture. The student may wonder, for instance, why the magnetic center of the planet does not coincide with the geometrical center. Such musings keep astronomy interesting.

Figure 5–5. *Spectacular Mercury photos from the Mariner 10 spacecraft on 29 March 1974. Top left: the southern hemisphere of the planet during approach. The spacecraft's distance was then 124,000 miles (200,000 kilometers). Top right: taken from 130,000 miles (210,000 kilometers), showing the other half of the illuminated hemisphere. Bottom: obtained from 3,700 miles (5,900 kilometers) distance, the area of photo measures 31 × 25 miles (50 × 40 kilometers). Craters 500 feet (150 meters) can be seen.*

Mercury has played a key role as a test object in the verification of Einstein's general relativity theory. In this theory, slight differences exist in predictions of motion in gravitation fields, in comparison to the predictions of the ordinary Newtonian law. Specifically, Einstein's theory predicted a slightly greater motion of the perihelion point of Mercury's elliptical orbit than that of Newton's theory. The latter held that the gravitational attraction of the other planets on Mercury, acting in addition to the major influence of the Sun, would cause the point of Mercury's closest approach to the Sun to advance slowly in the direction of its orbit. Einstein predicted a greater advance, but only a 43 arc-sec greater advance per century. This very small amount was actually found in the observations of the planet's motion, the observed number being 39.5 ± 5 arc-sec.

## Venus

Venus has a thickly clouded atmosphere. The atmosphere has frustrated all attempts to study the surface since Galileo's time. For many years, astronomers had little more to show for their study of Venus than estimates of its mass, size, and density, all of which closely resemble those of Earth.

Two early measurements of its temperature were available, and they differed substantially. From infrared observations, a value of 235°K was obtained; but the radio astronomers claimed that a body at 650°K would better fit their measures. Both studies could be consistent, however, if the infrared temperature referred to the cloudtops, while the radio temperature measure referred to the surface. The composition of the Venus atmosphere was known to consist mainly of carbon dioxide ($CO_2$) and a few traces of exotic compounds.

The information to this point is interesting in regard to the question of life on the planet. Plant life requires $CO_2$ and temperatures between those quoted above. Therefore, when it was decided to send probes to the vicinity of the planet, the probes were programmed to carry out measurements of the temperature and composition of the atmosphere. In 1962, the Mariner II spacecraft was launched for the purpose of determining the *surface* temperature of the planet. The results showed, conclusively, that the previous ground-based measures had been correctly interpreted as a genuine surface temperature, rather than, as some had suggested, a hot, tenuous outer layer of atmospheric gas. Now, if Earth were at Venus' distance from the sun, it would not be nearly so hot as 650°K. Therefore, it is certain that Venus' high surface temperature is due to its atmosphere acting as a blanket, efficiently retaining the energy pumped in by the sun. On Earth, such effects are seen in greenhouses, where glass allows sunlight to be absorbed inside, but traps the heat created. Venus has a marked greenhouse effect due principally to its $CO_2$ atmosphere.

The 1967 Mariner 5 spacecraft measured the temperature again and also made a clever remote chemical analysis of the atmosphere using a radio technique. A $CO_2$ concentration of 72 to 85 percent was found. Also in 1967, the Soviet interplanetary probe Venera 4 ejected a payload which parachuted toward the surface, measuring the temperature and pressure during its descent. These increased, respectively, to 580°K and 20 times our atmospheric pressure until the high pressure rendered the device inoperative. The payload also made direct chemical analysis of an atmospheric sample and found: carbon dioxide, 90–95 percent; nitrogen, under 7 percent; combined water and oxygen, 1.6 percent; water, 0.4 percent. Sometime later, another Venera craft reached the surface, reporting a temperature of about 650°K and an incredible pressure of 100 times the earth's surface. The surface of Venus is not a very hospitable place. Mariner 10 reached within 3,500 miles of Venus en route to Mercury, and clearly revealed a distinct weather pattern. While Venus itself is a slow rotator, its atmosphere swirls around the planet once in four days (Figure 5–6). An unexpectedly large shell of carbon monoxide (CO) has also been found to surround the planet.

*Venus' Atmospheric Mystery*

Why does Venus have such a thick mantle of $CO_2$? Strangely enough, Earth holds an answer to this question. On our planet, a very large $CO_2$ supply is locked up chemically in the rocks, mainly limestone. If this were released, the earth, too, would be similarly covered. Since water is the catalyst responsible for this locking-up process, Venus' large amount of $CO_2$ tells us that water is very scarce there. The lack of water on Venus is the real mystery.

*Radar Astronomy of Venus*

Venus, whose heavy cloud cover made its surface a mystery to the optical astronomer, has proved to be a most rewarding object for the radio astronomers. The diameter of the planet was found to be $12,112 \pm 1$ kilometers, or $7526 \pm 1$ miles. The first attempts to measure the rotation period seemed disappointing as very little Doppler shift was seen in the trailing parts of the returned radio signal, suggesting a long (many days) rotational period. The radio astronomers noted, however, that the

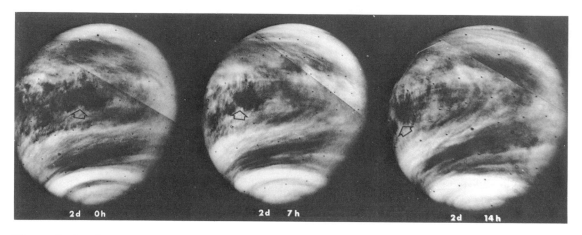

Figure 5–6. *Three views of cloud-shrouded Venus obtained by Mariner 10, which confirm a suspected four-day circulation of its atmosphere. The rapidity of this circulation for such a slowly rotating planet is unexpected. The clearly-delineated circulation pattern is in accord with theory, however. The photos reveal no surface features.*

planet had a highly reflective area which left an easily identifiable signature in the radio echo. From observing this area as it returned to view, the true rotation period was found to be about 243 days. Further refinement has confirmed this period and has shown that the direction of rotation is retrograde, with the equator tipped 102° away from the plane of the orbit. Most planets have their equators close to their orbit planes; the only other exception is Uranus.

The slow rotation of Venus is known to have a remarkable relation to Earth. This is shown in Figure 5–7. If we place an identifying mark on the surface element of Venus which is closest to Earth during an alignment (or conjunction), then when the next alignment occurs, we see this mark in the same place. We seem forced to conclude that the earth exerts a controlling influence on Venus' rotation. On the other hand, calculations show that any tidal effect on Venus due to the earth must be extremely small, probably too small to work any influence; very curious, indeed.

Finally, let us remark that the Mariner 5 spacecraft showed Venus' magnetic field to be virtually nonexistent. This is in accord with our ideas on rotational stirring of the core being the source of the field. On the other hand, without measuring a magnetic field, we cannot be sure if Venus has a liquid metallic core.

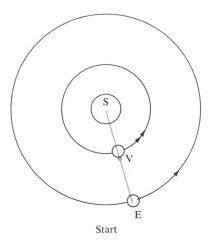
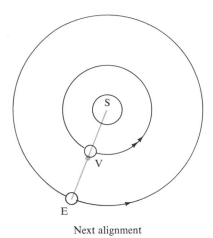

Start                              Next alignment

Figure 5–7. *The Venus-Earth relationship. A mark placed on Venus when the two planets are aligned is visible again on the successive alignments. Presumably, this results from some tidal effect due to the earth, although this is difficult to justify on theoretical grounds (see text).*

## Mars

Mars is probably one of the best-suited planets for telescopic study, because its atmosphere is usually transparent and it comes as close to Earth as 33,000,000 miles, or 53,000,000 km (Figure 5–8). Mars has two

Figure 5–8. *Three ground-based photos of Mars' Solus Lacus region, showing changes with time, due to wind-carried dust. A dust storm with winds up to 200 mph rages in the rightmost view. The effect of seasonal change on the south polar cap (top) is evident.*

diminutive moons, Phobos and Diemos, which give good estimates of the planet's mass using Kepler's third law. The size of Mars is directly measurable, and its rotation period, as well as the inclination of its pole, is readily deduced from its quite visible general surface features. Earth-based visual observations, which tend to reveal a little more detail than direct photographs, were the source of considerable controversy as they seemed to show long, thin, dark lines crossing the surface. Many years ago, some astronomers raised the question of these being canals dug by intelligent beings for the purpose of distributing water from Mars' readily visible polar cap. Now, however, the results of the Mariner space probe leave no doubt that the Earth-based observations suffered from some type of artifact probably related to the peculiarities of vision.

Earth-based observations, before the Mariner space probes, were able to determine the temperatures over the planetary surface and study the circulation patterns in the atmosphere by following cloud patterns. There are times when no detail can be seen. Winds, apparently whipping up surface dust, manage to obscure the whole planet for weeks at a time. The atmospheric composition was found to consist of a large measure of $CO_2$, at a pressure less than $1/10$ of the earth's atmosphere.

Since Mars' equator is tipped away from the plane of its orbit by $23°$ (very much like Earth's), it has seasons. These are twice as long as those on Earth because of the two-year revolution period. Mars' distance from the sun also varies by 20 percent due to its markedly elliptical orbit. This influences temperatures as well. Daytime temperatures as high as $70°F$ have been recorded near the equator. Effects of the seasons are seen clearly as advances and retreats of the polar caps. Seasonal changes are also seen in the general markings, but these effects are very likely not related to growth of living things.

As a result of the Mariner missions to Mars, we now have a very large number of high-quality photographs as well as spectroscopic and infrared measures covering a large fraction of the surface (Figures 5–9, 5–10, and 5–11). (Incidentally, the last two Mariner probes, 8 and 9, arrived during the height of a global dust storm.) As a result, not only is there considerably more insight into Mars as a geological entity, but certain information fundamental to an understanding of the early solar system has resulted.

The general picture emerging is that, geologically speaking, Mars lies somewhere between the Moon and Earth. The outer mantle, or lithosphere, of Mars is reckoned to be some 200 km thick. By contrast, the Moon's lithosphere is probably 300 km to 1,000 km deep, while the earth's is about 50 km. It is thought that the thickness of the lithosphere should be related in some way to the planet's geological activity. Volcanism, for instance, might be expected to be more important for thin-skinned bodies. Seismic data for the Moon and studies of Martian volcanoes seem to confirm this view.

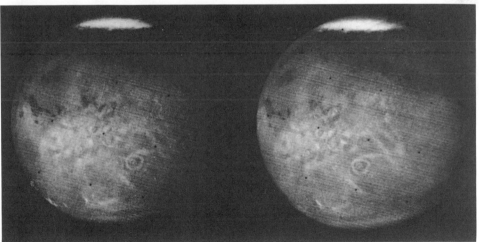

Figure 5–9. *Top: Photomosaic of the equatorial region assembled from Mariner 9 photos. Bottom: Mariner 7 photos from 281,000 miles (452,100 kilometers) at an interval of 47 minutes, showing the planet's rotation. The two photos are a stereoscopic pair; the three-dimensional effect can be obtained if 6 to 10-inch card is held vertically between the two photos and the head placed so that each eye sees one image only. The south polar cap is prominent. The bull's eye feature above center is the volcano Nix Olympica.*

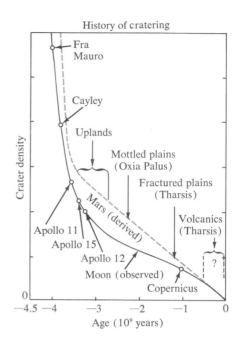

Figure 5–10. *The cratering history of Mars, compared to the Moon. The ages of several cratered lunar regions have been determined by dating Apollo rocks, deriving the curve shown in the figure. The ages of the cratered regions seen by the Mars probes are estimated, and give a similar curve. The similarity of the cratering history of these two bodies suggests that a common mechanism may have been at work to form the craters.*

One example of the implications of the Mariner photos for conditions of the early solar system is the study of crater densities on Mars. Mars' craters, like the Moon's, are mainly due to impacts, although there definitely are obvious examples of volcanic craters, such as the Nix Olympia feature (Figure 5–11). As Figure 5–12 shows, the older areas

Figure 5–11. *The giant (310 miles or 500 kilometers across) volcanic cone, Nix Olympica, seen by Mariner 9 as the great Martian dust storm, raging when the probe arrived, finally abated. Nix Olympica resembles Mauana Loa, the great volcano of the island of Hawaii.*

Figure 5–12. *Jumbled terrain. The break-up mechanism is not understood. The region measures approximately 200 miles (310 kilometers) across.*

of the Moon and Mars have been similarly pelted by interplanetary debris over the age of the solar system. Jupiter's gravitational influence on the minor bodies of the asteroid belt, the debris lying between the orbits of Mars and Jupiter, is assumed to be the cause of this pelting. Note that the ages of the crater areas on the Moon are known with certainty as a result of laboratory dating of rocks returned by the various Apollo missions, but the ages of the Mars areas must be *estimated* from the photos. Thus there seems to be little doubt that the inner planets of the solar system were heavily bombarded in their early history.

The Mars photos also indicate the possibility of water-ice lying in a thin layer over the frozen $CO_2$ polar caps (Figure 5–13). There also seems to be some strong evidence for the flow of liquid water on the surface at some time in Mars' history, according to interpretations of photos such as in Figures 5–14 and 5–15.

Even the two moons of Mars (Phobos and Diemos) were examined in the Mariner mission, and the region around the planet was searched for any new satellites. These curious bodies were first discovered about a century ago by Asaph Hall at the U.S. Naval Observatory. They are too small to show up as discs in earth-based photos, but their sizes were determined from their brightness—Phobos was estimated to be roughly 24 kilometers or 15 miles in size, and Diemos about half of that. The Mariner photos (Phobos is shown in Figure 5–16) clearly show them as objects 22 km and 12 km in diameter, respectively; Earth-based guesses were amazingly close. However, no Earth-based observatories could have achieved the Mariner's revelation that Phobos and Diemos are cratered and keep one face toward Mars like the Moon toward the Earth. Also like the Moon, these small bodies seem to be covered with a dusty layer, pulverized by impacting objects.

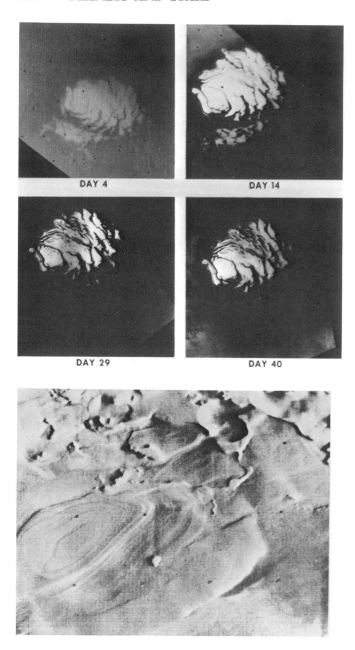

Figure 5–13. *Top: four views of the south polar cap, showing obvious changes in the frosty covering. Bottom: close up (29 × 37 miles, or 47 × 60 kilometers) shows a layered structure. The composition of the cap probably consists of carbon dioxide (dry ice) with the possibility of a thin covering of water-ice.*

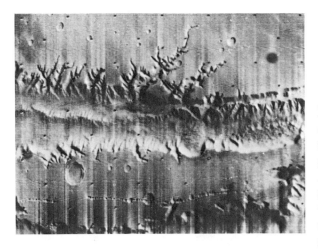

Figure 5–14. *A "Grander Canyon," 300 miles (480 kilometers) south of Mars' equator. The portion shown here is 235 miles (376 kilometers) long. This feature represents a type of landform evolution apparently unique to Mars. The streamlike nature of its "tributaries" is probably not an indication of water erosion, because many of the "tributaries" are closed depressions.*

The question of life in some form on Mars must still be left open. (Mariner's cameras could not have detected direct evidence of life *on the earth* from a similar altitude.) However, trace amounts of water vapor were detected on Mars, so we may be optimistic while awaiting the launch of an improved probe.

## Jupiter

After the Sun, Jupiter is the most massive object in the solar system. Although five astronomical units (a.u.) from the Sun, its great size makes

Figure 5–15. *Two vexing pictures. Both seem to indicate erosion by water, but water in the necessary quantity has not been found on the planet. Some water may be locked up in the polar caps.*

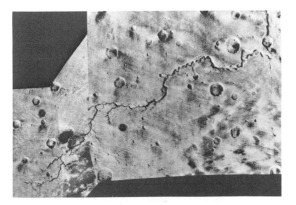

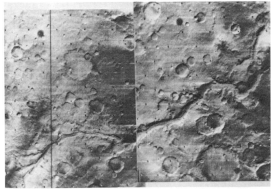

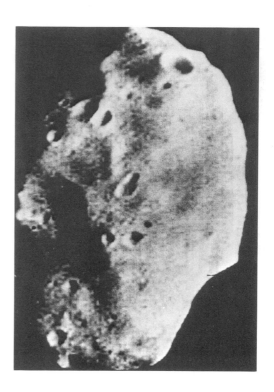

Figure 5–16. *Phobos, one of Mars' diminutive moons, photographed from a range of 3,400 miles (5,500 kilometers). Note the many small craters.*

it visible from the earth as a body of up to 40 arc-sec in diameter. Prominent are the belt-like features which parallel its equator and the strange Great Red Spot, visible since the mid-1800s. These can be seen in color plate 5, which if rotated 90° more readily shows that the planet's rapid rotation gives it a pronounced equatorial bulge. The surface features are not exactly permanent, because the visible surface does not rotate uniformly. The equatorial region rotates 5 minutes faster than the rest of the planet's $9^h55^m$. The visible surface is obviously not solid (Figure 5–17).

The low density of Jupiter shows that its composition is remarkably different from those planetary bodies we have already discussed. The lighter elements, notably hydrogen, must make up most of its bulk. In composition it is therefore much more like a star than a planet. Evidently, the central regions of the planet are not at a high enough pressure and therefore not at a high enough temperature to ignite the nuclear fuel. Studies of Jupiter in the infrared, conducted from the earth, and the more recent information from the Pioneer 10 spacecraft show that without doubt Jupiter returns more heat to the outside than it receives from the Sun. Thus, there is definitely an internal energy source present. Two explanations for this source seem possible: heating accompanying a slow contraction, or heating resulting from nuclear reactions deep in the

Figure 5–17. *Left: Jupiter, as seen by the Pioneer 10 spacecraft, at a distance of 1,-121,000 miles (1,804,000 kilometers) reveals a fascinating pattern of cloud swirls, below center in photo. Also noteworthy are the white spots which may be hot gas rising. The surrounding dark rings could be cool descending gas. Above: a close-up shows a wealth of detail, exemplified by delicate loops. They may play some role in the transfer of material along the planet's north-south direction. Limb of planet is at top.*

interior, like the stars, but on a very reduced scale. Despite the heat source, the measured outer temperature is a chilly 135°K.

In recent years a number of arguments have been raised in support of the possibility of life on Jupiter, or more properly, "in" its outer layers. The idea is that the temperature probably increases inward, and that all the chemical elements important for life are present, notably carbon, hydrogen, oxygen, and nitrogen. In addition, Jupiter has a very strong magnetic field, leading to a great deal of surface electrical activity, which has been studied from Earth by tuning to the radio waves it produces. This activity might supply the energy needed to form complex chemical compounds. Although somewhat speculative, the ideas are interesting.

Jupiter has the largest satellite system, including the famous four Galilean moons and eight other satellites. The outermost four satellites are extremely interesting because they revolve in retrograde fashion. It is conceivable that these are captured asteroids, because it is known that a larger body's capture of a small one is enhanced if the small one chances by on a retrograde passage.

### Saturn

Saturn is probably one of the most fascinating objects to view in a telescope due to its beautiful ring system (Figure 5–18). Structurally, Saturn
is similar to Jupiter, and it has the smallest density of any known planet,
even less than that of water. Like Jupiter, it rotates rapidly in $10^h38^m$,
and is even more bulged at its equator. It also shows non-uniform rotation of its visible surface. No evidence of internal heating has so far been
observed for this planet. Its surface temperature is 95°K.

Figure 5–18. *Saturn, photographed at different wavelengths, presents a different appearance. Note the banded structure of the
planet and the division in the rings, known
as Cassini's Division. The shadow of the
planet on the rings can also be seen.*

Its satellite system at first glance consists of ten bodies, the outermost of which is much more distant from the rest and revolves in retrograde orbit. Its innermost satellite, Janus, was just discovered in 1966. In
one respect, Saturn has the largest number of satellites of all the planets,
because its ring system must be composed of countless small orbiting
bodies. The ring system cannot be a solid disc because then the inner
and outer regions of the disc would revolve about Saturn in the same

period. This would be in violation of Kepler's third law relating the period and the size of an orbit; distant objects must have longer periods of rotation. The inner parts of the ring do indeed revolve faster than the outer parts. This has led to the proposal that the sizes of the ring particles must be very small, because collisions between bodies of different speeds would have worn them down like a gristmill.

### Uranus

On a very dark night, Uranus can just barely be seen by the unaided eye; thus the planet was the first to be discovered in relatively modern times. William Herschel discovered Uranus on March 13, 1781, as a greenish-appearing, nonstellar disk in his telescope. Very little is known of the surface of the planet as it is without distinctive markings; this follows from its small apparent size, about 4 arc-sec. Its internal construction is considered to resemble that of Jupiter.

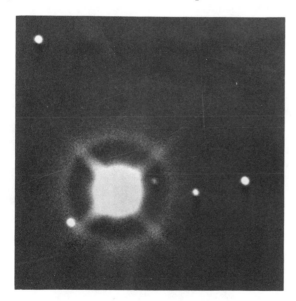

Figure 5–19. *Uranus, overexposed to show the satellites: Miranda, Ariel, Umbriel, and Oberon (in increasing distance from the planet). Very little detail is visible on the planetary disc under the best circumstances.*

Uranus has five satellites (Figure 5–19). Its most unusual characteristic is that its poles lie almost in the plane of its orbit. Since the polar axis maintains an approximately constant direction in space, this means that once during its revolution, each pole virtually points to the sun. It is an amusing task to consider the appearance of the sun during the $10^h49^m$-day as seen from Uranus, for different positions of the planet in its orbit. The surface of Uranus probably has a temperature less than $100°K$ and is frozen solid.

### Neptune

Neptune is not at all visible to the unaided eye. It was discovered by virtue of its gravitational disturbance on the newly-discovered Uranus, in what shall always be regarded as a triumph of dynamical astronomy. It was found on September 23, 1846, the first night an observer had the predicted position, and was within 52 arc-min of the predicted position.

Very little is known about Neptune's surface features, and like Uranus, its temperature is probably less than 100°K. Its internal structure resembles Jupiter's. It has two satellites: one close in retrograde revolution, and one distant in direct orbit. Its rotation period is thought to be about 15 hours.

### Pluto

Pluto is the most distant of the known planets. Because of its very small, even starlike, apparent size, we have only rough estimates of its size, mass, or density (Figure 5–20). The disturbance Pluto produced on the

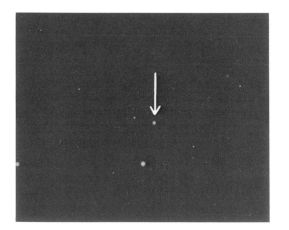 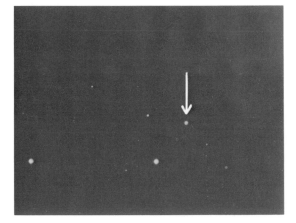

Figure 5–20. *The motion of distant Pluto seen in a twenty-four hour period. The planet is indistinguishable from a star in a photo.*

motion of Neptune was supposedly responsible for its discovery, in 1930, by Clyde Tombaugh. Lately, it has become apparent that this discovery was probably accidental and just good fortune.

Although Pluto barely presents a disc in the 200-inch telescope, we do know its period of rotation. This is because Pluto varies in light with a 6.4-day period, presumably due to an uneven reflectivity. The latest refinements in its dimensions are due to an interesting observational technique. The planet's path through the stars is calculated, and possible occultations of the background stars by the planet are noted. If an occultation occurs, its duration can reveal a lower limit for the planet's size, by assuming that the planet passed centrally across the star. One such possible occultation was monitored in 1965 and no eclipse occurred, indicating the planet's size to be less than 6,800 kilometers (4,200 miles) in diameter. Pluto has no known satellite. As a matter of fact, there is some debate as to whether Pluto could at one time have been a satellite of Neptune. Pluto does in fact have a highly elliptical orbit and it comes almost as close to the Sun as Neptune. This orbit is also markedly inclined to the majority of the other planetary orbits (by about 17°).

As of now, Pluto marks the outer planetary boundary of the solar system. A direct search for more distant objects would be extremely laborious, and perhaps of dubious value. Presently, it has a low priority among other tasks now commanding the attention of the planetary astronomer.

## Lesser Bodies

Comets and asteroids (or minor planets) represent only a negligible fraction of the total solar system's mass, but they are very important to an understanding of the solar system's evolution. It is important to stress that the asteroids bear a close relationship to the terrestrial planets, while the comets seem to be relics of a very early stage of solar system development. As will be noted, both comets and asteroids can manifest themselves as meteors, bodies which vaporize when entering the atmosphere.

### Asteroids

These "minor planets" comprise a band of tens of thousands of small bodies in rather circular orbit between Mars and Jupiter. In 1801, an Italian astronomer, Giuseppi Piazzi, discovered the first asteroid, Ceres, which at 480 miles in diameter is also the largest. Five asteroids are larger than 100 miles in size, but the typical diameter of a minor planet may be less than a mile.

The asteroids are considered to be another example, like Saturn's rings, of collisions wearing bodies down to ever-smaller dimensions. Some of these collisions can apparently impose radical changes in the orbits of the objects involved—we find a small group of asteroidal bodies in *elliptical* orbits, coming closer to the sun than the orbit of the earth. This observation lends weight to the idea that debris which evidently pelted the Moon, Mars, Mercury, and presumably Venus as well, came from the asteroid belt. Thus, when we put large meteorites in our museums, we are very likely handling interplanetary material which once called home an orbit at 2.8 a.u. from the sun (Figure 5–21).

The rain of particles falling onto the earth still continues; we can see it for ourselves with the unaided eye on a clear, moonless night. With a bit of patience, one will perceive a *meteor*. On most evenings, meteors will occur with random directions, and these are called *sporadic meteors*. On the other hand, several times a year, the number of meteors will be well above average, with the trails appearing to diverge from a particular point on the sky. This is a "meteor shower" and will be discussed later.

A meteor is a trail of hot atmospheric gases left behind by a bit of interplanetary junk, typically the size of a pea, as it vaporizes in the atmosphere (like re-entry of a spacecraft). Meteors, of course, are the objects commonly referred to as "shooting stars." Astronomers have added a bit to the confusion, perhaps, by calling the particles in space *meteoroids,* and when landed (if not completely vaporized) *meteorites.*

It is further interesting to mention that those living away from cities can have their own collection of meteoroids, provided one is satisfied with particles of a size visible in a microscope. Every day, many thousands of tons of these lightweights, known as *micrometeorites,* fall to the earth's surface after being gently slowed in the upper atmosphere. The particles may be gathered by placing trays on a roof and sorting the debris after a few months, using a magnet. Under a microscope, micrometeorites have a quite jagged appearance, readily distinguishing them from other natural particles. The technique works in cities, but it may be difficult to differentiate the micrometeorites from industrial pollutants.

## Comets

Comets are perhaps the most visually spectacular component of the solar system, although they comprise very little of its mass: .0001 percent (Figure 5–22). Most comets travel in very long elliptical orbits, reaching the vicinity of the outer planets. Comets are definitely members of the solar system. Orbit calculations for some, however, show that they left the Sun on paths such that they would never return. When these comets were studied more carefully, it became evident that either Jupiter or Saturn had gravitationally influenced them previous to their passing close to the Sun. Furthermore, before that influence, the calculations showed that the orbits had been elliptical; thus these bodies, too, belonged to the Sun.

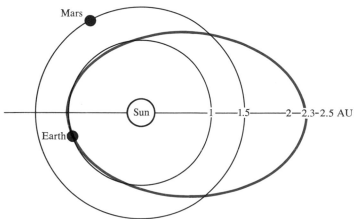

Figure 5–21. *Top: the Lost City, Oklahoma, meteorite, recovered on 9 January 1970. On 3 January, it was photographed by the Prairie Network of meteor-patrol cameras. Bottom: the orbit calculated from the trail photographs, taken at different locations. Note that the orbit reaches to the distance of the asteroid belt.*

Before dealing with the question of the origin of the comets, let us digress to consider what we know of their nature. A comet must be typically less than a trillionth of Earth's mass. This was discerned when one was observed passing through the region of Jupiter's moons, and no

Figure 5–22. *Photo of the naked-eye comet, Comet Bennett, taken at the Goddard Space Flight Center on 16 April 1970. The comet could be easily seen in the early morning hours with the unaided eye, and made a fine object for binoculars.*

gravitational influence on them was noted. From its behavior as it approaches the Sun, and from the results of spectroscopy, there is a good general picture of a comet as a "dirty ice-ball" roughly 1/2 mile in diameter. At distances exceeding Jupiter's, a comet has an almost starlike appearance. On passing inside the orbit of Jupiter, however, gases begin to evaporate from the surface of the comet and form a cloud, or *coma*, about the starlike nucleus. Following this, a tail (or tails) develop, generally pointing away from the Sun's direction. This kind of activity usually grows in intensity as the comet approaches close to the Sun (Figures 5-23 and 5-24). At perihelion, it may break up due to tidal effects of the Sun on it and/or some material may be vaporized. Very probably each passage removes about 1 percent of the body's mass. Apparently, periodic comets do not last very long!

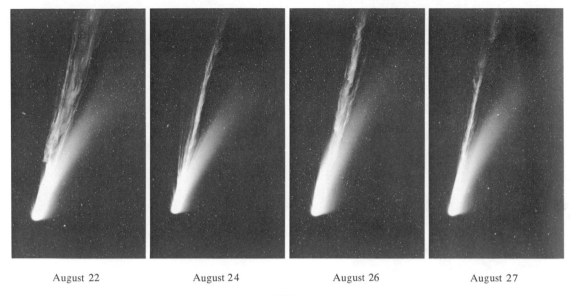

August 22          August 24          August 26          August 27

1957

Figure 5–23. *Four views of Comet Mrkos, 1957, showing marked changes in one of the tails. The constant, diffused tail is shaped by the pressure of sunlight, whereas the turbulent tail is due to an interaction of the comet's ionized gases with the interplanetary magnetic field and the solar wind of corpuscular radiation. (The solar wind represents particles which the sun constantly boils off into space and which consist primarily of hydrogen.)*

Here we meet an interesting problem. We have two known ways to lose comets: the influence of Jupiter and Saturn on their orbits, and their vaporization by successive perihelion passages. It therefore behooves us to ask how the comet supply is replenished. The answer seems closely connected with the origin of the solar system. The Dutch astronomer J. Oort suggests that the comet source is a spherical cloud of cometlike bodies at a very great distance, perhaps 100 a.u., from the Sun. At that distance, the circular velocity is very low (keep in mind that a body at rest falls in toward the Sun). Thus it may happen that gravitational

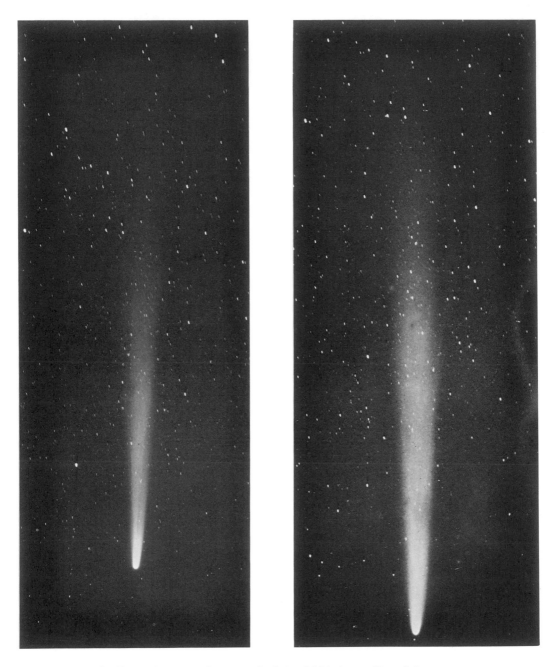

Figure 5–24. *Halley's Comet, photographed in 1910 from Honolulu. The longer tail is 40° in length. The earth passed through the tail of the comet (a very tenuous medium, an excellent vacuum on earth) and will do so again in 1986.*

forces from the nearby stars can influence the bodies to fall in, where they may be seen as comets.

As noted, comets are on occasion observed to break up at perihelion, usually into two parts, which continue on in very similar orbits. It is known that many comets "die" by shredding themselves in this fashion until matter is strewn all along the orbit, uniform for some and clumped for others. The phenomenon of meteor "showers," seasonal displays of meteors, are evidence of this. For example, suppose the orbit of a comet crossed that of the earth [the situation of Figure 5–21(b)]. Both bodies would not likely be at the intersection simultaneously. However, if cometary debris became strewn all along an Earth-intersecting orbit, then it would be quite possible that a collision with the swarm of debris would occur. If the pieces were very small, they would burn up quickly in the atmosphere. Figure 5–25 shows that such particles would be traveling

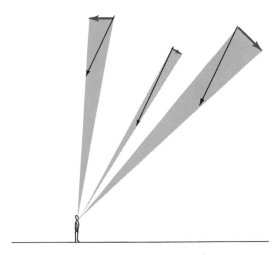

Figure 5–25. *Parallel meteor trails appear to diverge, as seen from the ground. The trails' apparent intersection on the sky marks the shower radiant. The black lines represent the actual parallel trails of the meteors, while the blue lines represent the apparent trails.*

along together, and in parallel paths. When we on Earth see a meteor "shower," and plot the meteor trails on a sky may, it quickly becomes evident that the trails appear to diverge from some point, or at least a small region, on the sky. Such a point is called the *radiant* of the meteor shower. The convergence of the trails is just a perspective effect, akin to staring down a street whose parallel sides seem to merge in the distance. So it is that the parallel trails of the co-moving particles appear to converge. The constellation containing the radiant gives its name to the shower. Meteor showers associated with a particular intersecting orbit are *annual* occurrences, as the intersection with the stream lies at a certain point in the *earth's* orbit. From the yearly differences observed in the number of shower meteors, we can actually infer something about the uniformity or clumpiness of the cometary debris in its orbit. The dates and names of some major meteor showers are given in Table 5–2; the absence of a bright moon on those dates will facilitate observation.

### TABLE 5–2

*Major Meteor Showers*

| Constellation of radiant | Date of best display | Related comet |
|---|---|---|
| Quadrantid | 3 January | |
| Lyrid | 21 April | 1861 I |
| Eta Aquarid | 4 May | Halley* |
| Delta Aquarid | 30 July | |
| Perseid | 11 August | 1862 III |
| Draconid | 9 October | Giaccobini-Zinner |
| Orionid | 20 October | Halley* |
| Taurid | 31 October | Encke |
| Andromedid | 14 November | Biela |
| Leonid | 16 November | 1866 I |
| Geminid | 13 December | |

* Crosses earth's orbit twice.

## Formation of the Solar System

The oldest rocks on the surface of Earth have been dated by a reliable method which uses radioactive decay of the element uranium into the element lead. Theory gives the rate at which the decay occurs; thus by studying the *ratio* of lead to uranium in certain rocks, the time span over which the process has acted can be determined. Results give $4.6 \pm 0.1$ billion years. This age is very close to the probable age of the Sun, so it seems likely that the planetary system was formed at about the same time that the Sun was formed. Therefore, it is very likely that both were formed in the same process. Since the Sun is in fact a typical star, it must have formed as other stars do; namely, by a contraction and condensation of gases and dust already present in the interstellar medium. Now, the interstellar medium has a density of only about 1 atom per $cm^3$, so a *volume* of material occupying a much greater dimension than the current solar system must have been originally involved.

A natural feature of the contraction of such a mass is that it must spin faster during collapse (just as a spinning ice skater rotates more rapidly by contracting his arms). Also, regarding the mass as initially spherical, we realize that there will be a greater tendency for the poles, which are not spinning, to fall inward faster than the rapidly rotating equator region. We can imagine the early solar system as a kind of disc. What we can term the *proto-sun* would look like a bulge in the middle of the disk and be spinning rapidly. When the size of this disc was close to the diameter of the present planetary system, the planets were presumably forming from condensations in the flattened part, called the *solar nebula*. The nebula by this time had probably already formed small particles; these may have stuck together on impact to build up larger bodies. In order to understand how the colliding particles could stick, it has been suggested that they were covered with a layer of ice.

At present, we do not have an entirely clear picture of events following this stage, but some imagination can carry us further. It seems likely that the *terrestrial* planets began as bodies more like Jupiter's (and the Sun's) composition and that their current scarcity of lighter materials, such as hydrogen and helium, probably resulted from heating by the early Sun. This heating also drove off debris left between the planets and swept it clear of the planetary system. Could this be the source of Professor's Oort's proposed comet cloud? It does seem possible, because the best model we have for a comet is the dirty ice-ball.

Questions regarding the origin of the satellite systems of the planets and the asteroid belt are even more difficult to answer. However, it does seem likely that satellite systems, such as the directly-revolving moons of Jupiter, could have formed concurrently with the parent planet, analogous to the planets with the Sun, whereas the retrograde moons may be captured bodies.

Earth may have gained its satellite by adoption or by birth; that is, either by the capture of an already-formed body, or directly from the earth itself by a splitting process known as fission during an early, hot, fluidlike phase. Fission could perhaps occur if the earth's rotation were very rapid, too rapid for a simple flattened disk to be stable. We have to picture here a spinning droplet splitting into two unequal parts. It is not certain whether this process can actually work for a planetary body. We may note, however, that the overall composition of the moon resembles the earth's upper layers. Nevertheless, this only tells us that the moon *might* have been part of the earth.

The capture theory is equally uncertain. If the moon were captured, it would probably have been captured from a retrograde orbit, close to Earth. Some elementary physics shows that the earth's rotation must then have been about three hours. Tremendous tides would have been caused on the early Earth, perhaps enough to melt most of it in a short (astronomically speaking) time. There *is* evidence that the earth's core was formed in a rapid process, interestingly enough. Conversion of the

Moon's orbit to a direct one would have been accomplished by the inclination of its retrograde orbit increasing, until it passed through a polar orbit and then became direct. It would have then moved slowly away from earth.

We see that the problems connected with the formation and evolution of the solar system are a rich field for research. A reasonably complete understanding will be no simple task.

## Questions

1. What are the two major solar system groups? How are the major planets divided?

2. Why are Mercury and Venus difficult telescopic objects? How may Venus be best studied?

3. What suggests that some nuclear reactions are going on in Jupiter's interior?

4. Why might it not be worthwhile to search for a tenth planet?

5. What are the asteroids? How are they related to the craters on Mercury, the Moon, Earth, and Mars?

6. What general evidence suggests that the planets and the sun were formed at the same time?

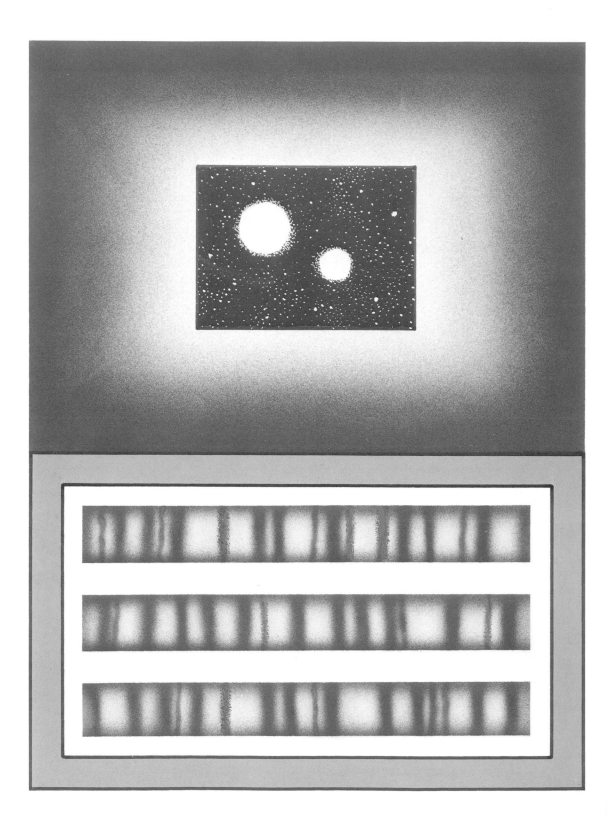

# 6

## Basic and Observable Properties of Stars

### Introduction

Our understanding of the stars—what they are, how they are formed, what changes they undergo during their existence, and how they end—is ultimately based on *estimates* of one or more basic properties of a relatively small number of stars. These basic properties are:

$L$ = luminosity (amount of energy of all kinds radiated per second)

$M$ = mass (amount of material making up the star)

$R$ = radius (size of the star)

$T_S$ = surface temperature

[E/H] = surface chemical composition; that is, number of atoms of chemical element E relative to number of atoms of hydrogen, the simplest element and by far the most abundant in ordinary stars

Notice that these properties only tell us about the outside of the star; they tell us nothing about the inside. How can we be sure that two stars having the same outside properties are identically arranged inside? The truth is, we cannot. Since we cannot see inside, we have to use the laws of physics to predict what conditions exist inside the star. The predictions based on these laws (called "stellar models," to be discussed later) indicate that, for a given set of values of $L$, $M$, $R$, $T_S$, and [E/H], there is only one arrangement of the star's interior satisfying the known physical laws.

Our understanding of the stars is based on estimates of the basic properties because in general they are not directly observable. In fact, much effort is exerted in determining these estimates. There are, however, some observable aspects of a star. A perceptive observer of the night sky should notice several obvious yet significant features of these tiny points of light. First, some appear brighter than others. Second, there are suggestions of color about them. For example, in the constellation of Orion, prominent in the winter sky, there is a bright star that seems reddish and another at the other end of the constellation that appears blue-white. (See Figure 6–1.) Third, the stars are not distributed at random. For example, not too far from Orion, in the constellation of Taurus, there is a very compact group of fairly bright stars known as the Pleiades (Figure 6–2). The latter is an example of a *cluster* of stars.

Figure 6–1. *Below: the constellation of Orion (the Hunter), prominent in the winter sky in the Northern Hemisphere, in a long-exposure photograph. Right: map of Orion indicating the colors of some of the brighter stars.*

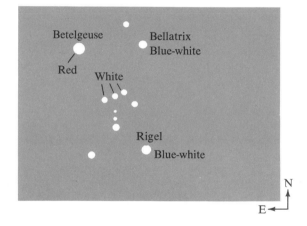

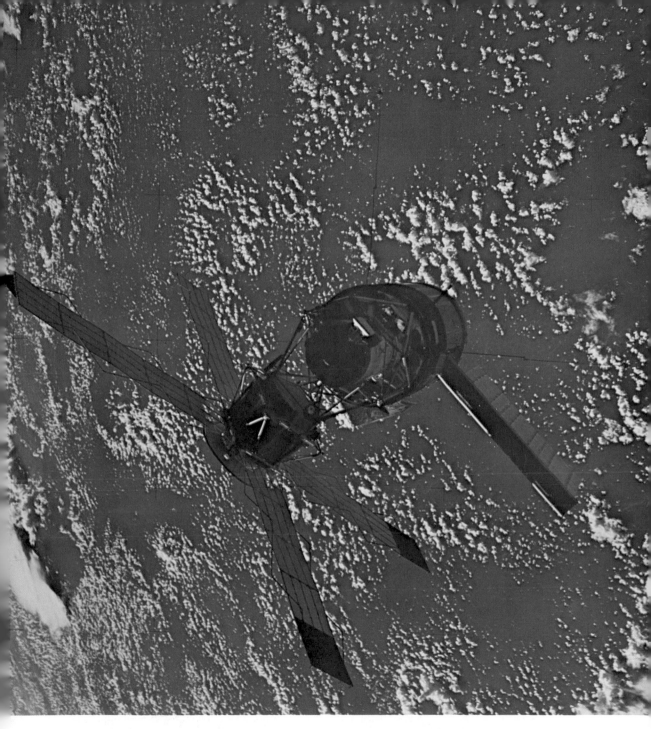

The Skylab in earth orbit seen against clouds and sea. The four solar panels in x-form belong to the Apollo Telescope Mount. One of the orbital workshop's panels has been severed in an accident. A record series of physical, biological and physiological experiments were performed. (Plate 1)

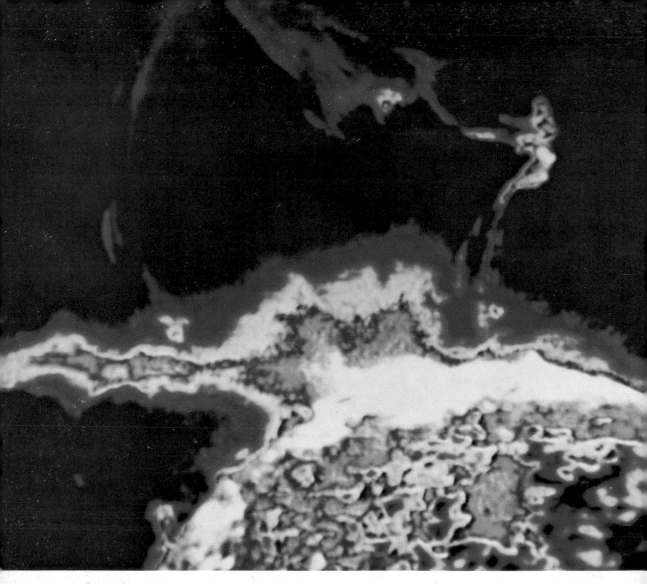

**Skyab 3 photo of a vast solar prominence taken in ultraviolet light. Intensities of light in the photograph are coded into colors yielding this dramatic view. (Plate 2)**

A photo of Comet Kohoutek taken by the Sky-lab astronauts on Christmas Day, 1973. The pressure of sunlight has blown back the dust and gas given off by the comet. The photo was taken with ultraviolet light, but the intensity of ultraviolet light has been color-coded to yield this striking photo. (Plate 3)

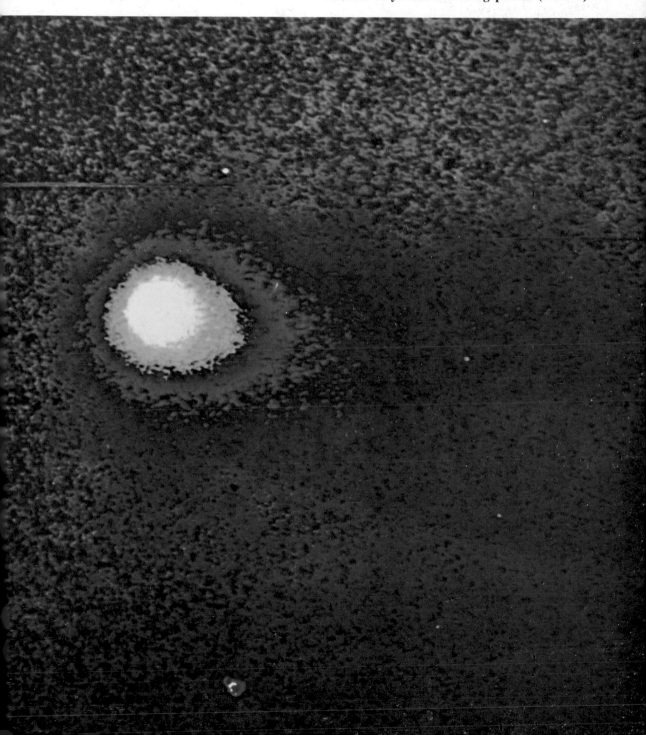

Scientist-astronaut Harrison H. Schmitt surveys a huge boulder near the Taurus-Littrow landing site of the Apollo 17 mission, the last lunar landing. Eugene A. Cernan took the photograph, while Ronald E. Evans flew the mother ship in lunar orbit. (Plate 4)

Pioneer 10 viewed Jupiter from 1,615,000 miles (2,600,000 km) in this photo. The shadow of the innermost satellite Io is seen on the disc, and the Great Red Spot is visible. In 1976 Pioneer 10 crosses the orbit of Saturn. (Plate 5)

The Earth gleams brightly against the stark, black backdrop of space as photographed by the Apollo 16 astronauts during their Earth-Moon round trip. The United States and other parts of North America are clearly visible in this photograph. (Plate 6)

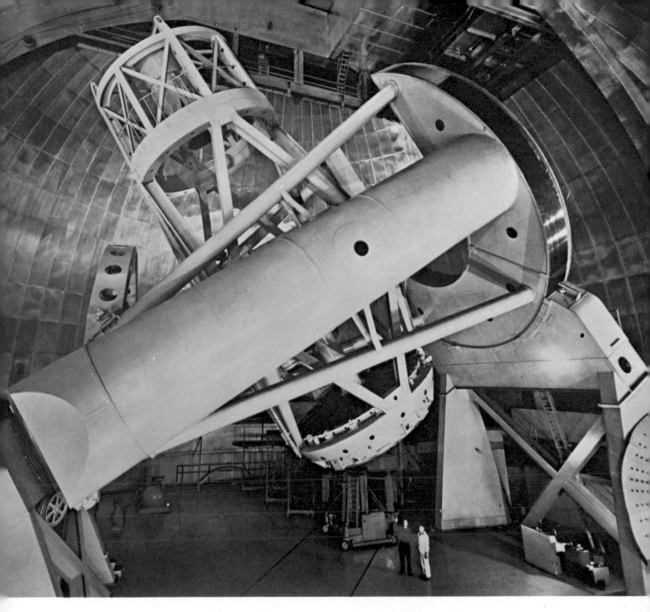

**General view of the 200-inch Hale telescope from the east. (Plate 7)**

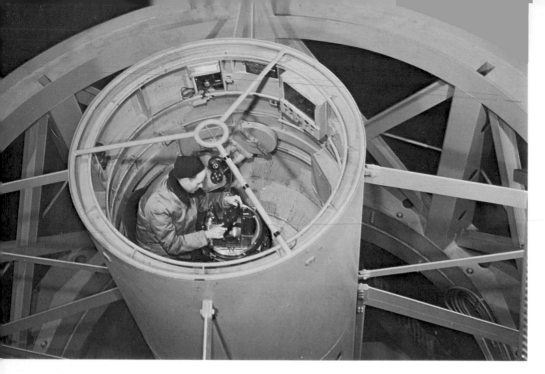

**View of top of 200-inch Hale telescope with astronomer in the prime focus cage. (Plate 8)**

**Interior of the Coudé room of the 200-inch Hale telescope. (Plate 9)**

The Pleiades and nebulosity in the Taurus constellation; NGC 1432, Messier 45. 48-inch Schmidt photograph. (Plate 10)

The Trifid Nebula in the Sagittarius constellation; NGC 6514, Messier 20. 200-inch photograph. (Plate 11)

**The Horsehead Nebula in the Orion constellation; NGC 2024. 48-inch Schmidt photograph. (Plate 12)**

**The Great Nebula in the Orion constellation; NGC 1976, Messier 42. 200-inch photograph. (Plate 13)**

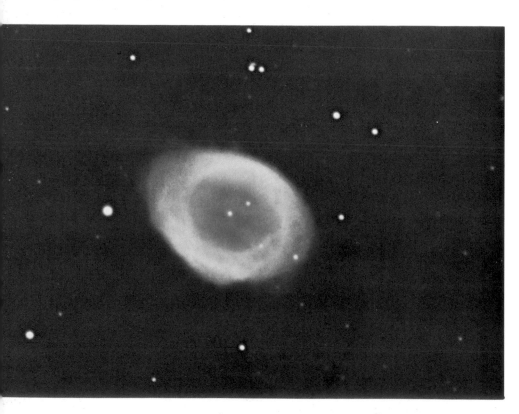

**The Ring Nebula in the Lyra constellation; NGC 6720, Messier 57. 200-inch photograph. (Plate 14)**

The Dumb-bell Nebula in the Vulpecula constellation; NGC 6853, Messier 27. 200-inch
photograph. (Plate 15)

The Crab Nebula in the Taurus constellation; NGC 1952, Messier 1. 200-inch photograph. (Plate 16)

The Great Galaxy in the Andromeda constellation; NGC 224, Messier 31, 48-inch Schmidt photograph. (Plate 17)

An irregular galaxy in the Ursa Major constellation; NGC 3034, Messier 82. 200-inch photograph. (Plate 18)

Figure 6–2. *The Pleiades, a star cluster in the constellation of Taurus (the Bull), which can be viewed in the evening during autumn in the Northern Hemisphere. At first this loose group of stars may look like a hazy spot to the unaided eye; with good eyesight and a dark sky, six stars may be distinguished. With low-power binoculars fifty or more stars may be seen. Note nebulosity around the stars.*

(See also color plate 10.) Many times one will see two stars very close together; these are *double stars* or *binaries* (Figure 6–3). In addition to these classes of objects, there is a hazy band of light stretching in a great arc across the sky called the Milky Way. Closer examination of this hazy glow reveals that it is a collection of millions of faint stars in dense groupings, like clouds of stars. Later, when we discuss galaxies, the significance of the Milky Way will be made clear.

Figure 6–3. *Krüger 60B, a binary star. Note the orbital motion of the companion in the sequence of photographs.*

Some deductions can be made from these rather simple observations. The different apparent brightnesses of stars could imply: (1) that some stars are actually brighter than others, that is, they have greater *absolute* brightness; (2) that some are more distant than others; or (3) that some of the stars have greater or smaller absolute brightness *and* are more or less distant than others. In most cases, (3) is the most probable. The stars' different colors can be related to the different colors of an ideal radiating body (chapter 3). We remember that a red-hot object is cooler than a white-hot object. The stars do not appear to be distributed at random, which indicates that there is some large-scale structure to the system of stars, and it seems most reasonable to suppose that the double stars and star clusters are held together by the gravitational forces exerted by their members on each other. This is in fact the case.

## Stellar Spectral Types

### Classification Scheme

If we had to treat each star as a special case, we would make little or no headway in understanding them because of the great number of stars, so we look for similarities and patterns among them just as a botanist or zoologist does with plants or animals. In order to find similarities and to distinguish one kind of star from another, we employ a spectrograph when possible (see Chapter 4). The analysis of starlight by spreading it into a spectrum with a spectrograph reveals the great variety of stars. The spectra in Figure 6–4 demonstrate the differences among types. Just as Sherlock Holmes could identify the brand of a cigar from the ashes it left, an expert can immediately classify a star by looking at the distinctive patterns of dark lines (*absorption lines*) or

Figure 6–4. *The principal types of normal stellar spectra. Important absorption lines used in classifying the spectra are indicated. The chemical symbol for the element giving rise to the particular line is followed by a Roman numeral which indicates what state of ionization of the atom gives rise to that line. The designation "I" refers to the atom with the usual number of electrons, i.e., the same number of electrons as protons (hence electrically* neutral); *the designation "II" to that atom with one electron missing, i.e.,* singly ionized; *the designation "III" to that atom with two electrons missing or* doubly ionized, *and so forth.*

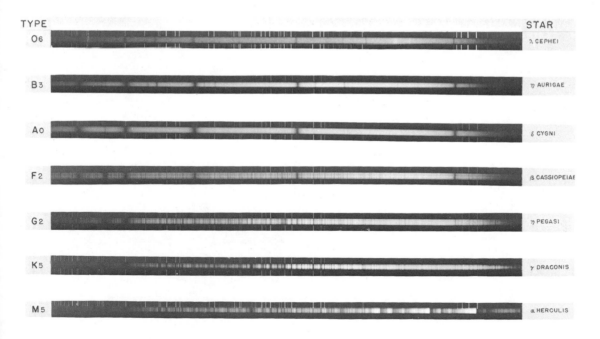

bright lines (*emission lines*) in its spectrum. The *spectral types* widely used in astronomy to classify the majority of stars form a sequence along which the strengths of the characteristic absorption lines due to the different elements increase or decrease smoothly from one class to the next. The colors of the stars also vary smoothly along the sequence (Table 6–1).

As the variation in color would suggest, the sequence is one of *decreasing* surface temperature. We will see later that the orderly variation in the strengths of the different spectral lines along the sequence comes about for the same reason. The letters for the different types are not in

Figure 6–5. *Strengths of absorption lines of various elements for different temperatures and spectral types. The variations in strength are due to changes in the excitation and ionization of the atoms with temperature, as discussed in Chapter 3. The differences for the different chemical elements arise because some of them are easier to ionize than others. For instance, calcium (CaI) requires relatively little energy (and thus a relatively low temperature) to become ionized, hydrogen (HI) requires more energy (and a higher temperature), and helium (HeI) very much more energy (and much higher temperature).*

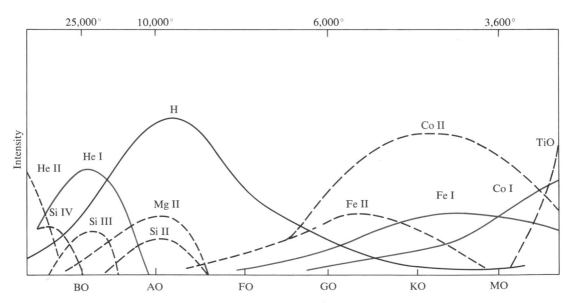

## TABLE 6–1

*Spectral Classification*

| Type | Color |
|------|-------|
| O | blue-white |
| B | |
| A | white |
| F | yellow-white |
| G | yellow |
| K | orange |
| M | red |

alphabetical order because the sequence was originally arranged in order of decreasing strength of the spectral lines due to the element hydrogen. Fortunately there is a traditional mnemonic for committing the sequence to memory: *"Oh, be a fine girl, kiss me!"** 

## Luminosity Classes

It is possible and, as a matter of fact, necessary to classify stellar spectra with higher precision than that of the seven main classes listed in Table 6–1. First, each letter class has been divided into ten parts or subclasses. A star whose spectrum looks intermediate between a purely B-type spectrum (B0) and a pure A-type spectrum (A0) is classified as B5. The sun is assigned a spectral type of G2 on this system, which means its spectrum resembles that of a G0 star more closely than that of K0 star. Second, it is known (by means that we will go into shortly) that some stars of a given spectral type are much larger and more luminous intrinsically than others *of the same type.* They can be identified in many cases simply from the appearance of their spectra, for example, from the peculiar sharpness of their spectral lines compared to those of their smaller counterparts. Depending upon how much more luminous these stars are than the others of the same spectral class, they are referred to as *subgiants, giants, bright giants, or supergiants.* The faint members of each spectral type, which for spectral type G2 includes the sun, are referred to as *dwarfs.* The categories of luminosity are referred to as *luminosity classes* (Table 6–2) and are assigned Roman-numeral designations (to avoid confusion with the alphabetic spectral types and the Arabic-numeral subclasses).

---

* Attributed to the American astrophysicist Henry Norris Russell.

## TABLE 6–2

*Luminosity Classes*

| | |
|---|---|
| Ia | Bright supergiants |
| Ib | Supergiants |
| II | Bright giants |
| III | Giants |
| IV | Subgiants |
| V | Dwarfs |

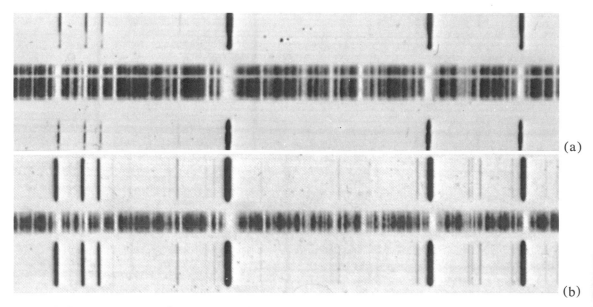

(a)

(b)

Figure 6–6. *Luminosity effects on stellar spectra. (a) Section of a spectrogram of the Sun, classified as spectral type G2V. (b) Section of a spectrogram of the giant star η Herculis, classified as spectral type G4III. Note how the Sun's absorption lines appear broader than the corresponding lines of η Herculis. The stellar spectrum in the middle is flanked by a comparison spectrum above and below.*

### Special Classes

In addition to these categories, there are a number of more specialized designations. There are a few stars with spectra somewhat resembling the

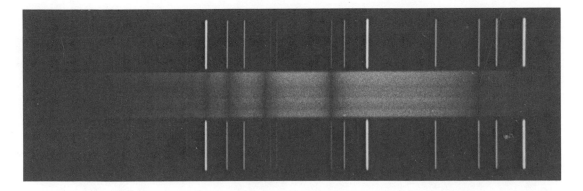

Figure 6–7. *The spectrum of a white dwarf, BD+73°8031. The absorption lines in a white dwarf are decidedly broader than those of a main sequence star of the same temperature. The broadening is due to the high pressure in the white dwarf's atmosphere.*

spectra of A- and B-type dwarfs but with unusually broad spectral lines. In addition, these stars are much fainter than ordinary A stars. They are called "white dwarfs," and they are a special type of star that we will discuss later (Chapter 9), when we see how stars evolve. The white dwarfs have their own system of spectral types, with several categories like the ordinary categories but with the prefix D, for example, DA, DB. Some have spectra that are completely without lines, that is, *continuous* spectra. These are classified DC.

There are also stars whose spectra show very broad emission lines instead of absorption lines. This type of spectrum was first observed by the French astronomers Wolf and Rayet in the last part of the nineteenth century. Stars with this kind of spectrum are referred to as Wolf-Rayet stars. Generally this is abbreviated WR, although the more specialized notations WC and WN are used. (The C refers to the emission bands being predominantly due to the element carbon, whose chemical symbol is C, while the N refers to nitrogen.) Otherwise normal stars which show a few narrow emission lines simply have the letter "e" added to their spectral type. An example would be B2Ve, which means a B2 dwarf with emission lines in its spectrum. The emission usually occurs with very hot stars (types O and B) or unusual relatively cool stars (type M). Stars which have the absorption lines of a particular element (usually some metallic element, although in the star's atmosphere it exists as a vapor) unusually strong or weak also have a special designation. For reasons that are still not understood, these peculiar line strengths occur almost entirely within spectral type A.

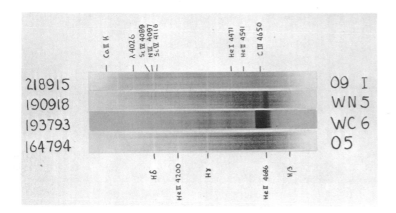

Figure 6–8. *The spectra of two Wolf-Rayet stars shown beside spectra of two normal O stars. The Wolf-Rayet stars have spectra that are basically similar to those of the O stars except for the characteristic broad emission bands. These bands are actually spectral lines seen in emission because they come from gases* around *the star, not in the star's inner atmosphere. The great width of the lines is caused by a Doppler shift; the surrounding gasses are streaming away from the star at high velocity (as much as 1000 kilometers per second).*

By far the majority of stars fit into the ordinary spectral types, and, as we mentioned above, there is a correlation of color with spectral type. This suggests that we can estimate a star's spectral type simply by measuring its color, and to a certain extent this is true. (The word "color" has a well-defined meaning in astronomy, as described in Chapter 4.) As we shall see, however, the colors are affected by the material between the stars through which the starlight travels.

## Binary Stars

### Kinds of Stellar Systems

The membership of a star in a double star system or any other star system is not really an observable quantity in the same sense as spectral type, apparent brightness, or color, but it is important. There

Figure 6–9. *NGC 2682 (M67), a galactic (open) cluster in the constellation of Cancer (the Crab). Another well-known galactic cluster, the Pleiades, is shown in Figure 6–2.*

are pairs of stars close together *(binary stars)*; systems with several stars together *(multiple star systems)*; systems with a few hundred stars *(galactic star clusters)*; systems which are somewhat similar to galactic clusters but are much more loosely connected *(associations)*; systems which contain hundreds of thousands of stars and are much more tightly packed than galactic clusters *(globular clusters)*; and larger assemblies of stars, millions to hundreds of billions *(galaxies)*. We will discuss galaxies in Part 3. If we can establish that a star belongs to one of these systems and can somehow find out how far away that system is, we will immediately know the distance of the star in question. Of course this is true only if the size of the system is fairly small compared to its distance from us. It is not true for the nearest star clusters, for example, but in almost all other cases it is true.

## Types of Binary Stars

Much of our knowledge of the basic properties of stars comes from studies of binary stars, so we should outline the various kinds of binaries. By coincidence some stars appear close together, while in reality they are

Figure 6–10. *NGC 6205 (M13), a globular cluster in the constellation of Hercules. Note how many more stars it has than the galactic clusters and how much more regular it appears. The stars in the center are not as tightly packed as they seem. Because of atmospheric "seeing" and unavoidable photographic effects, the images of stars overlap and run together at the center of the cluster.*

quite distant from each other. Since these *optical pairs* (as they are called) are not gravitationally bound to each other, we can extract no useful information from them. Other close pairs *(visual binaries)* can be seen to orbit around each other if we have measurements of their positions relative to each other over a sufficiently long period of time. If we measure their positions with respect to a number of background or *reference* stars, we can locate the center-of-mass around which both are moving. There are cases where a second star cannot be seen; only the motion of one star around the center-of-mass is visible. This motion shows up as a very tiny "wiggle" in the star's movement across the sky and is detectable only by very careful measurements of the star's position or (as it is called) *astrometry*. This type of binary is referred to as an *astrometric binary* (Figure 6–11).

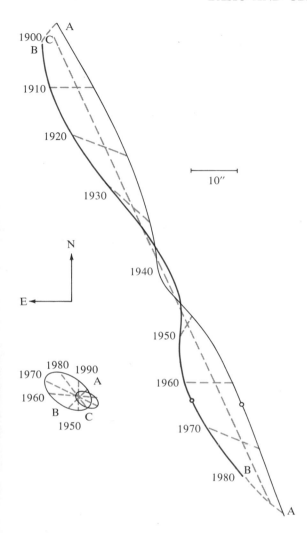

Figure 6–11. *Sirius as an example of an astrometric binary. It was noticed in the mid-nineteenth century that Sirius' motion across the sky was not straight but had a very small "wiggle," presumably due to a companion star that was much fainter than Sirius and thus could not be seen in the glare from the latter. Some decades later, when because of its orbital motion the companion was not so close to Sirius in the sky, it was finally observed.*

A star may also appear to be single in the largest telescopes, but if its radial velocity (see Chapter 3) changes in a certain periodic fashion, we can tell that we are actually looking at two stars—a *spectroscopic binary* (since the radial velocity is measured on spectrograms). If sets of spectral lines from both stars can be seen in the combined spectrum of the pair and the radial velocity of each star measured, the system is a *double-line spectroscopic binary* (Figure 6–12). If only one set of spectral lines can be seen, the system is a *single-line spectroscopic binary*. If both sets of lines can be seen but no radial-velocity changes caused by the stars' orbital motion are apparent, the system is called a *spectrum binary*. A star whose apparent brightness changes is called a *variable star*. There are several different kinds: *pulsating variables,* whose brightness changes

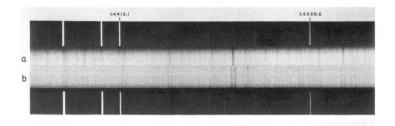

Figure 6–12. *Spectrograms of Mizar, the middle star in the handle of the Big Dipper (constellation Ursa Major), at two different times. This star is an example of a double-line spectroscopic binary; note the two sets of lines. The spectral type is A2 and the period is 20.5 days.*

as they rhythmically expand and contract;* *eruptive variables,* whose brightness increases dramatically and then slowly decreases; and *eclipsing variables,* in which we are interested here. Eclipsing variables or eclipsing binaries are pairs of stars in which one star passes alternately in front of and behind the other, eclipsing each other so that the light from one of the stars is partially or totally blocked from our view (see Figure 6–13). The variations in brightness caused by eclipses are in general very different from those due to other causes.

## Observational Selection Effects

These different categories of binary stars are by no means mutually exclusive. Some binary stars are both visual and spectroscopic binaries, which means we can measure their motion on the sky and the variations in their respective radial velocities as well. Other binaries are both eclipsing and spectroscopic. If this were not so, we would have great difficulty in obtaining estimates of masses and radii for stars. It is, however, extremely rare for a system to be both eclipsing and a visual binary. There is a good reason for this. In order for us to see stars eclipse one another, one of two things must occur: (1) we must either be in or near the plane of the stars' orbital motions, or (2) the stars must be very close together; that is, their separation must be comparable to or less than their radii (Figure 6–14). The chances of our being in

---

* Pulsating variables show periodic changes in radial velocity because of their pulsation, but these changes are different in form from orbital variations.

Figure 6–13. *Superimposed photographs of WW Cygni, an eclipsing variable star, taken at maximum and minimum brightness. The images of surrounding stars do not differ substantially in the two photographs.*

the orbital plane or sufficiently close to it to see the eclipses are very small. For the visual binaries, in order to resolve the binary into two stars, the two must be separated by at least 0.5 arc-sec because of the limited resolving power of the telescope (Chapter 4). At the typical distance of a star, this angle corresponds to a separation much larger than the radius of an average star. We see, then, that the eclipsing binaries generally have small separations, while visual binaries are usually widely separated. If we look at a histogram of the separations (Figure 6–15), we see that this is the case.

By Kepler's third law (Chapter 2), the orbital speeds will be greater on the average for close binaries than for wide binaries. The larger orbital speeds will be easier to detect spectroscopically, so spectroscopic binaries should have preferentially smaller separations than visual binaries, on the average.

Figure 6–14. *Illustration of the problem in discovering binaries that are both eclipsing and visual. (a) If the separation is large, the system can be resolved in a telescope as a visual binary, but the chance of our being located in just the right spot to see an eclipse is tiny. (b) If the separation is small, eclipses are fairly likely to be observed, but the two stars cannot be resolved as a visual binary at the typical distance of a star from the sun.*

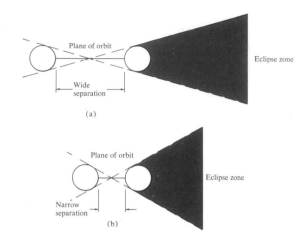

Figure 6–15. *Histogram of semi-major axes of random samples of eclipsing spectroscopic and of visual binary stars. Note how the former have mainly small separations, while the latter have large separations.*

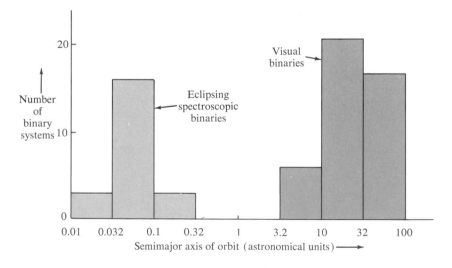

These properties are simple examples of *selection effects:* our techniques of observation and/or the capabilities of our equipment determine to a certain extent what values of the measured quantities will be obtained. If we were to study only visual binaries, for example, and determine the sizes of their orbits, we might falsely conclude that binaries with separations less than a certain value never occur in nature. Much effort in astronomy is directed towards eliminating this kind of bias from the data.

## Questions

1. Which of the following are basic properties of a star?
   (a)   mass
   (b)   apparent brightness
   (c)   color
   (d)   types of absorption lines in the star's spectrum
   (e)   radius
   (f)   membership in a binary-star system

2. Which of the following has the *higher* surface temperature? Why?
   (a)   Star A: reddish appearance in sky, spectral type 09
   (b)   Star B: yellowish appearance, spectral type G0

3. What kind of star would have spectral type M3III (color, size)?

4. Consider two lists of stars: the first a list of the twenty nearest stars, and the second a list of the twenty stars that appear brightest in the sky (smallest apparent magnitude). List 1 is comprised almost entirely of low-luminosity red stars, while list 2 contains a large fraction of high-luminosity blue stars. (Remember, luminosity refers to how bright the star really is, not merely how bright it seems in the sky in relation to other stars.) Does the fact that list 2 has many high-luminosity stars suggest a selection effect? What does the difference between the two lists tell us about the relative numbers of high- and low-luminosity stars in a given volume of space?

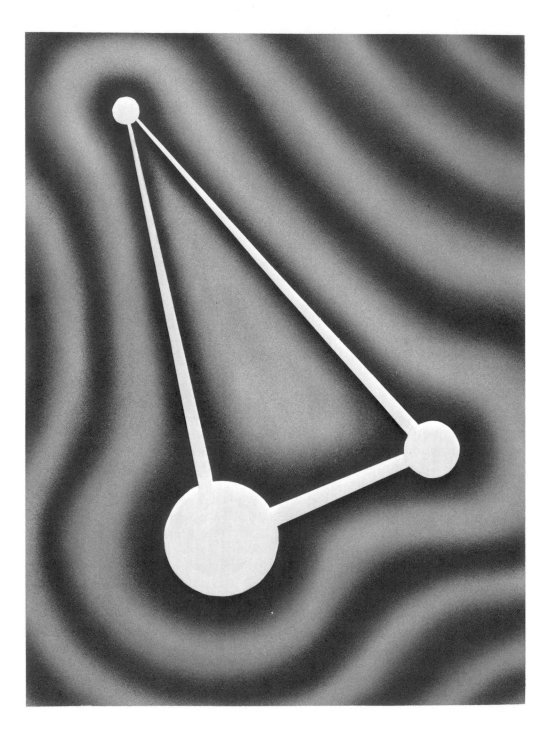

# 7
# Determining the Distances of Stars

## The Importance of Stellar Distances

In estimating the basic properties of stars, we usually need to know their distances. Without knowing how far away a star is, we cannot tell whether it is nearby and very faint (low luminosity) or far away and extremely bright (high luminosity). As we will see in Chapter 8, the estimates of mass and radius usually depend on the distance also. For these reasons distances are very important. Not only are they very difficult to obtain with accuracy, but distance values for one type of star often depend on the values determined for another type (or types). If a pivotal value has to be changed, as happens occasionally, the consequences can be far-reaching. We must understand that the system of astronomical distances is a structure carefully pieced together from many diverse clues, making use of much of astronomy.

We can consider the ways of finding distances to be of two types: *direct methods,* where the distance is found geometrically with a minimum of assumptions, and *indirect methods,* where fairly strong assumptions are made about the brightnesses or motions of stars. It need not be the case that an indirect method is inferior to any direct method; they all have their limitations, as well as circumstances under which they work well.

## Direct Methods

### Trigonometric Parallax

All direct methods involve astrometric measurements—measurements of the positions of objects on the sky—which we discussed in Chapter 4. There are a number of ways of finding distances directly, but they are essentially the same in principle as the primary one, the method of *trigonometric parallax*. Binoculars, stereo viewers, stereophonic sound systems—all these make use of the same basic principle as the method of trigonometric parallax. This same principle and our having two eyes instead of one enable us to judge distances. Let us consider the diagram in Figure 7–1. The lines from *A* and *B* to each of the points *P, Q,* and *R* represent the directions each of our eyes would be looking. We see that the farther away a point is, the more nearly alike the two directions are. In other words, the angle *APB* becomes smaller as *P* is moved farther away. Thus, angle *AQB* is smaller than angle *APB*, and angle *ARB* is still smaller.

How can we tell this difference in direction? A person's brain must be able to tell the direction each of the eyes is pointing (not consciously, of course!) and estimate from that how far away the point is. We can estimate distances without a background, but it is easier to estimate this difference in direction with the help of more distant objects in the background. Since the more distant objects differ less in direction than a nearby object as seen from two distinct locations, they furnish a frame of reference within which we can measure the difference in direction of

Figure 7–1. *Change in angle with distance. The difference in direction of a point seen from two separate locations is smaller as the point is more distant.*

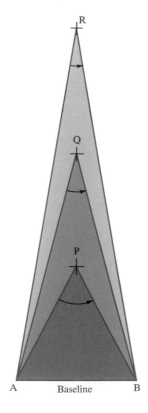

the latter. The application of this idea to stars is shown schematically in Figure 7–2. Points $A$ and $B$ represent the earth's position at two different times, six months apart (on opposite sides of its orbit). Points $X$ and $Y$ represent stars much more distant than star $P$. The directions to these stars define angles on the sky; for example, the directions to $X$ and $Y$ from $A$ form the sides of angle $XAY$. If $X$ and $Y$ are sufficiently far away, the directions to $X$ and $Y$ from $B$ are essentially the same as those from $A$. Suppose we take a photograph of the field of view containing $X$, $Y$, and $P$ when the earth is at $A$ and then another photograph six months later, when the earth is at $B$. If we lay one of the photographs over the other (assuming that they are transparent) so that the images of $X$ and

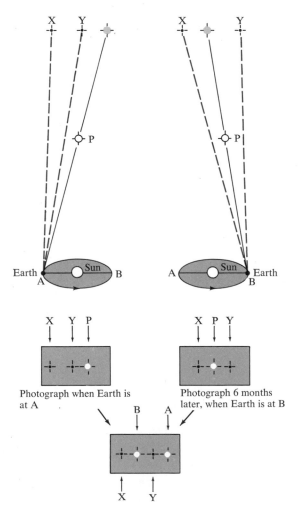

Figure 7–2. *Trigonometric parallax. Star P shifts position relative to the more distant stars (X,Y) as the earth orbits around the sun. The amount of the shift—parallax—depends on how far away star P is. The shift can be seen if photographs taken six months apart are compared.*

Photograph when Earth is at A

Photograph 6 months later, when Earth is at B

Composite of the two photographs

*Y* are aligned, the images of *P* would not be aligned. The angle of shift of *P* on the sky is very nearly the same as the angle *APB*. If the stars *X* and *Y* were infinitely far away, these two angles would be identical. Notice that the two photographs correspond exactly to the two views in the stereoscopic viewer (Figure 7–1). If we could superimpose those, we would see the largest difference in position on the slides for foreground objects and the smallest difference for background objects.

The method of trigonometric parallax always uses the diameter of the earth's orbit as the distance $\overline{AB}$. In actually calculating stellar distances, however, we use one-half the angle *APB*, which is the apparent size of 1 a.u. (the *radius* of the earth's orbit) as seen from the star. Half of angle *APB* is called the parallax *p*, and it is given in arc-seconds. Because this method is so fundamental to astronomy, a natural unit of distance (instead of miles or even astronomical units, where the numbers are inconveniently large) is that distance from which an astronomical unit would appear 1 arc-sec across. This unit is termed a *parsec*, because it is the distance corresponding to a *par*allax of 1 arc-*sec*ond.

How far is a parsec? We first notice (Figure 7–3) that, for a very small parallax *p*, the length of the chord *AC* is approximately the same fraction of the circumference as the parallax is of 360° (a full circle).

Figure 7–3. *Relationship between parallax and distance. The parallax is the angle* P *taken up by the radius of the earth's orbit—1 astronomical unit—at the distance of the star. If the distance* d *were larger, the angle* P *would be smaller.*

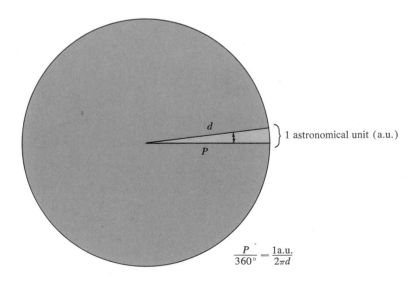

1 astronomical unit (a.u.)

$$\frac{P}{360°} = \frac{1\,\text{a.u.}}{2\pi d}$$

This gives us the proportionality:

$$\frac{1 \text{ a.u.}}{2\pi d} = \frac{p}{360°} \tag{7-1}$$

where $d$ is the distance from the star to the sun (and earth). We have to convert 360° to arc seconds, which is 360° × 60 arc-min per degree × 60 arc-sec per arc-min = 1,296,000 arc-sec. Then, if $p = 1$ arc-sec,

$$d = \frac{1 \text{ a.u.}}{2\pi} \times \frac{360°}{p} = \frac{1 \text{ a.u.}}{2\pi} \times \frac{1,296,000 \text{ arc-sec}}{1 \text{ arc-sec}} = 206,265 \text{ a.u.} \tag{7-2}$$

Once we know the size of the astronomical unit, we can find the length of the parsec. It turns out that a parsec is 3.26 times as far as light can travel in a year moving at 300,000 kilometers per second (186,000 miles per second). We observe from formula (7–2) that

$$d = \frac{1}{p} \text{ parsecs} \tag{7-3}$$

So the smaller $p$ is, the greater the distance. If the parallax is 0.01 arc-sec, the distance must be $(1/0.01) = 100$ parsecs.

As we have discussed it, the method has assumed that the background stars do not shift direction and the solar system and star are not really moving between photographs. This is certainly not the case. There are small corrections that must be made for the parallax shifts of the background stars, and by making measurements over a number of years we can distinguish the star's own motion from its parallactic shift.

There is an important (and unfortunate) limitation of this method. We have seen that the parallax is smaller the more distant the star. When the true parallax is smaller than the inevitable errors of measurement, the distance estimate is worthless. With present techniques, 100 parsecs is about the greatest distance to which we can use the method.

## Other Direct Methods

If we have some way of finding out how fast a star is moving at right angles, to its direction from earth, its *tangential velocity* (so called because it is at a *tangent* or right angle to the line of sight; see Figure 7–4), and if we can measure how fast it appears to move across the sky, we are able to calculate its distance in the same way as with the trigonometric parallax method. The apparent angular motion across the sky caused by the star's own motion in space is called its *proper motion,* denoted by the Greek letter $\mu$. It is called the star's "proper" or "own" motion to distinguish it from apparent motions such as the parallax shift.

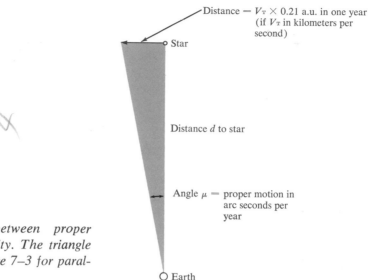

Distance = $V_T$ × 0.21 a.u. in one year
(if $V_T$ in kilometers per
second)

○ Star

Distance $d$ to star

Angle $\mu$ = proper motion in
arc seconds per
year

Figure 7–4. *Relationship between proper motion and tangential velocity. The triangle is much like the one in Figure 7–3 for parallax.*

○ Earth

It is usually expressed in units of arc-seconds per year and is measured in much the same way as a parallax, as a shift in position on the sky relative to more distant reference stars. However, fairly long time intervals (at least a decade) are necessary to measure it accurately for any star.

To find the distance, we first convert the star's speed into astronomical units per year. For example, the speed will commonly be given in kilometers per second. If we use the facts that the earth's orbital speed is 30 kilometers per second, the circumference of its orbit is $2\pi$ a.u., and the time for a complete revolution is 1 year, we find that

$$30 \text{ kilometers per second} = 2\pi \text{ a.u. per year}$$

or

$$1 \text{ kilometer per second} = 0.21 \text{ a.u. per year}$$

If the star's speed at right angles to its direction from earth is $V_T$ kilometers per second, it moves $0.21 \times V_T$ a.u. in a year. From the formula we used for trigonometric parallax, we see that

$$d = \frac{0.21 \times V_T}{\mu} \text{ parsecs} \qquad (7\text{–}4)$$

There are only special situations in which we will know the speed $V_T$. If we can determine in what direction in space a star cluster is moving (which we can do for the nearest clusters) and measure its radial velocity from the Doppler shift (Chapter 3), we can correct for the projection effect and calculate its tangential velocity (see Figure 7–5).

(a)

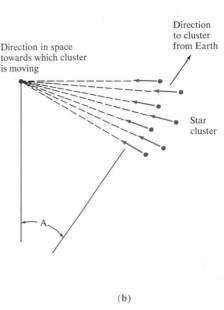

Direction in space
towards which cluster
is moving

Direction
to cluster
from Earth

Star
cluster

A

(b)

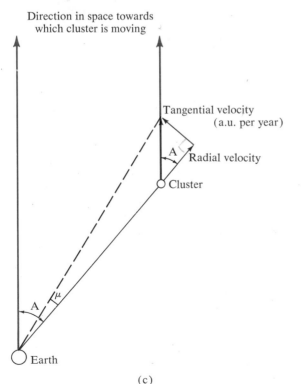

Direction in space towards
which cluster is moving

Tangential velocity
(a.u. per year)

A

Radial velocity

Cluster

A    μ

Earth

(c)

Figure 7–5. *Distance determination for a nearby star cluster. (a) The convergent-point effect. Although the telephone line runs parallel to the road, the two seem to converge at a point in the far distance. (b) The convergent-point effect for stars in a nearby cluster. The cluster stars are generally moving parallel through space (except for their random individual motions, which are small), so they appear to be converging on a point in the sky. We can trace the proper motions forward on the sky to find the convergent point, then measure the angle A on the sky between the convergent point and the cluster. (c) The angle A is the angle between the direction of the cluster's motion and the line of sight from earth to the cluster. If we measure the cluster's radial velocity (from Doppler shifts in the spectra of cluster stars), we can use trigonometry to find the tangential velocity of the cluster from angle A and the radial velocity. If the cluster's proper motion μ and tangential velocity are known, we can find its distance.*

Or if a shell of gas is expanding, we can measure its speed of expansion. Usually the gas is transparent, and we can detect the Doppler shift of light emitted by both the side of the shell towards us (blue-shifted) and the side away from us (red-shifted). This is illustrated in Figure 7–6. Presumably the speed of expansion is roughly the same in all directions, and we can, in principle, measure the proper motion of the shell's edge relative to the center. In general, however, we do not know $V_T$, and we have to use its average over a number of stars in place of its actual value for a given star.

There are errors of measurement with proper motions as well as with parallaxes. However, no matter how small the star's proper motion, with enough time (and patience!) we can measure it. This is not true with the parallax; the star simply appears to move from one side to the other of its true position in the sky. The effect of the star's proper motion accumulates over long periods of time—if it is as small as 0.001 arc-sec per year, after a century the star will have moved 0.1 arc-sec, a comparatively easy angle to measure. The use of photography to determine accurate positions has been around less than a century, so we cannot yet measure such small motions accurately. But it is interesting to speculate on how accurate proper motions could be a century from now even without the inevitable improvements in technique. At the present, the distance-determination methods using proper motions are only rarely useful to distances of 1,000 parsecs or more (and then only when $V_T$ is unusually large). Commonly they are useful to at most a few hundred parsecs.

## Indirect Methods

### Spectroscopic Parallax

The direct methods we have just discussed have essentially involved judging distances on the basis of how large something of known size appears. This can be compared to telling the distances of automobiles at night by how close together each car's headlights appear (Figure 7–7). We have a sense of how far apart they really are, and by noticing how close together they appear, we can tell whether the car is nearby or far away. But at great distances we cannot see the headlights separately, and we estimate the distance of a car by how bright its headlights appear. The dimmer they are, the farther away the car must be. Of course this does not work if the car's headlights are much dimmer or brighter than the lights of most cars. Also, if there is heavy fog, the

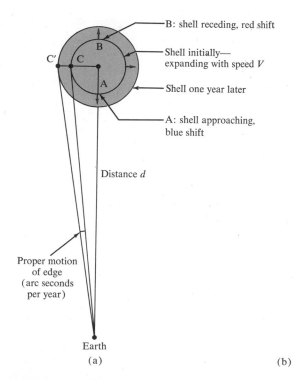

B: shell receding, red shift

Shell initially—
expanding with speed $V$

Shell one year later

A: shell approaching,
blue shift

Distance $d$

Proper motion
of edge
(arc seconds
per year)

Earth
(a)

(b)

Figure 7–6. *The expanding-shell method. The proper motion of the edge of an expanding gas shell can be measured, either by measurement of its position relative to a star in the middle or by taking half of the increase in apparent diameter. The actual speed of expansion* V *is one-half the difference in radial velocity between point A and point B. Usually the shell is transparent, and at the center of the shell the spectrum shows two sets of emission lines. The one shifted to shorter wavelengths corresponds to point A, where the gas is coming towards us, and the one at longer wavelengths corresponds to point B, where the gas is moving away. The photo shows an actual case in which the method was used: Nova Persei, 1901.*

headlights will seem very dim or even be invisible when the car is relatively close. Usually, however, this way of estimating distances serves us well enough.

Figure 7–7. *Headlights on a highway at night. The closer together the headlights appear, the more distant the car is. The most distant cars are too far away to be seen having two headlights; they appear as one. The more distant the car, the fainter the headlights appear.*

The same approach is used for stars, and the most basic of the indirect methods is based on it. This is the method of *spectroscopic parallax.* To understand how it works, we first remember that the area of a sphere of radius $r$ is given by the formula

$$A = 4\pi r_2^2 \qquad (7\text{–}5)$$

If a source of light is located at the center of the sphere and emits an amount of light $E$ in one second, each unit of area on the sphere receives the amount of light:

$$I = \frac{E}{A} = \frac{E}{4\pi r^2} \qquad (7\text{–}6)$$

This is the inverse-square law for light, illustrated in Figure 7–8. The larger the sphere (a greater $r$), the less light each unit of area receives. This means that if two identical telescopes collect light from the same source, one at a distance $r_1$ and the other at distance $r_2$, the amount of light they receive will be in the ratio

$$\frac{I_1}{I_2} = \frac{E}{4\pi r_1^2} \times \frac{1}{\frac{E}{4\pi r_2^2}} = \frac{E}{4\pi r_1^2} \times \frac{4\pi r_2^2}{E} = \frac{r_2^2}{r_1^2} \qquad (7\text{–}7)$$

We also see that by measuring the amount of light received, we can calculate how much light the source emits (if we know its distance):

$$E = I \times 4\pi r^2 \qquad (7\text{–}8)$$

In Chapter 4 we saw that astronomers use the system of magnitudes to indicate the amount of light received. The amount of light received per unit area corresponds to the apparent brightness, so we refer to *apparent magnitudes* of stars. The total amount of light emitted by the source is the absolute brightness, which we refer to in terms of *absolute magnitude.* The absolute magnitude of a star is defined as the apparent magnitude a star would have at the standard distance of 10 parsecs.

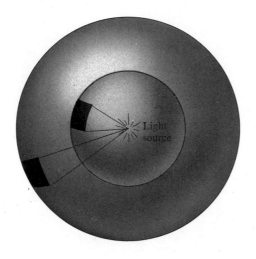

Figure 7–8. *Inverse-square law for light. The two patches have equal areas. But because the same amount of light must pass through both spheres, a smaller fraction of it passes through the patch on the larger sphere.*

From the definition of the magnitude system and the inverse-square law of light we have the important relation

$$m - M = 5 \log r - 5 \qquad (7\text{–}9)$$

where $m$ is the star's apparent magnitude, $M$ is its absolute magnitude, and $r$ is its distance in parsecs. The quantity $(m - M)$ is usually referred to as the *distance modulus*. The farther away the star, the larger $m$ is, which simply means that it appears fainter. Also, the more distant the star, the larger its distance modulus is (Figure 7–9).

If we measure the star's apparent magnitude and determine its distance, we can calculate from Equation 7–9 what its absolute magnitude must be. (We assume that its distance has been found by trigonometric parallax or some other method besides spectroscopic parallax.) Usually we can assume that other stars of the same type (identified by their spectra, for example) will have the same absolute magnitude. (This absolute magnitude is the brightness of our "standard headlight.") Then from this absolute magnitude and the apparent magnitude of each star, we can find the distance of each. This is a very powerful method of finding distances, adequate over hundreds of thousands of parsecs. On the other hand, if the original distance is incorrect, the absolute magnitude will be wrong, along with the distance values for the other stars of that type.

**Interstellar Absorption of Starlight**

In the example of automobile headlights, we considered the possibility of fog dimming the light and making the automobile appear farther away

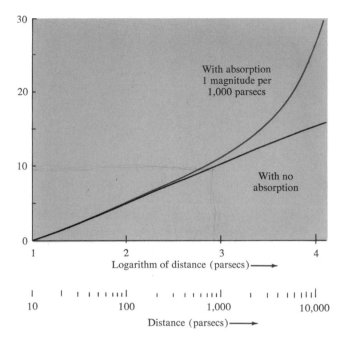

Figure 7–9. *Dependence of distance modulus on distance. The black line shows the relationship if there is no absorption, and the blue curve shows the relationship if there is uniform absorption of 1 magnitude per 1000 parsecs. Note that the distance has been plotted on a* logarithmic *scale, so that each large scale unit represents a factor of 10.*

than it really is. There is no fog in space—at least, not dense clouds of water vapor—but there is enough dust between the stars to dim their light appreciably. With the method of spectroscopic parallax, interstellar absorption of starlight by dust makes the stars seem farther away. For this reason we have to correct for the dimming of starlight, either by finding out how much absorption there is or by assuming that there are a certain number of magnitudes of absorption per 1,000 parsecs (see Figure 7–9). If the amount of absorption per *parsec* is $A$ magnitudes, the fundamental equation becomes

$$m - M = 5 \log r - 5 + Ar \qquad (7\text{-}10)$$

One way of finding out the amount of absorption is by taking stars with known absolute magnitudes and distances. If these stars are in the

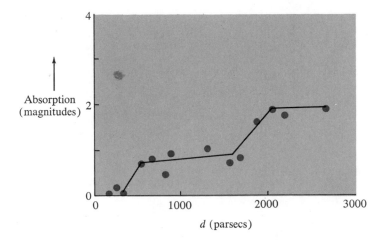

Figure 7–10. *Dependence of absorption on distance (schematic). The plotted points represent absorption values found for stars whose distance and absolute magnitude are known. The solid curve is an estimate of the absorption as a function of distance.*

same direction as the star we are interested in and are at different distances, we can determine the amount of absorption as a function of distance (see Figure 7–10). Another way involves the effect of interstellar dust on starlight. Suppose we see distant stars whose spectra are of type O. They should be bluish stars, but their light appears reddish. The dust scatters the star's blue light without greatly affecting the red light, in the same way that molecules and dust in the earth's atmosphere scatter the blue light from the sun and pass the red, giving us red sunsets. Careful measurements of the colors of stars will reveal how much the light is reddened as compared to the normal colors for that spectral type. The amount of absorption has been found to be proportional to the amount of reddening (in magnitudes) on the average, so we can calculate the former. Then, as before, we use the equation

$$m - M = 5 \log r - 5 + \text{abs (absorption)} \qquad (7\text{–}11)$$

### Variable Stars and the Period-Absolute Magnitude Relation

Certain types of *variable stars* can be readily identified by the pattern of their variation, notably the RR Lyrae and Cepheid variable stars. (These types are named for the stars RR Lyrae and δ Cephei, the prototypes of the respective classes.) Typical light variations are shown in Figure 7–11.

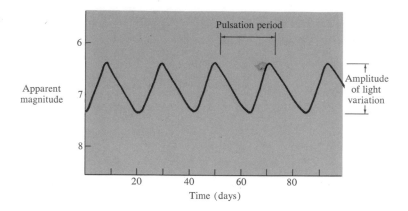

Figure 7–11. *Light variation for a type of pulsating variable star (a* Cepheid). *The period is the time interval between consecutive occurrences of maximum (or minimum) brightness.*

It has been found that there is a relationship between the period of variation for these stars and the absolute magnitude. As we see in Figure 7–12, this relationship is not exact, but it is nonetheless extremely helpful in estimating absolute magnitudes and, from absolute magnitudes, distances for individual stars can be determined. (Again, it is necessary to include absorption.) These are not the only types of variable stars; there is a wide variety, and by no means do all vary in brightness in a regular fashion.

### Statistical Parallax and the Solar Motion

An important indirect method which is used to find the absolute magnitudes of stars of the different spectral types—our "standard headlights" —is the *statistical parallax.* The method is similar to the direct methods discussed earlier that compared the velocity at right angles to the line of sight with the proper motion. The difference lies in the fact that, as the name implies, the statistical parallax compares an *average* velocity with an *average* proper motion.

The sun and its neighbors are all moving more or less together around the center of our galaxy. In addition to this general motion of the solar neighborhood as a whole, the sun has a slight motion of its own in space. This is known as the *solar motion,* or the sun's *peculiar* (that is, nonsystematic) *velocity.* Each of the sun's neighbors has its own peculiar

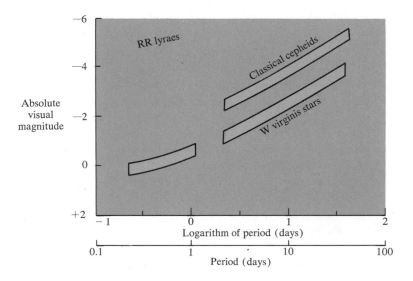

Figure 7–12. *Relationship between period and mean absolute visual magnitude for pulsating variable stars. The relation is not precise, so it is represented by a band instead of a curve in the figure.*

velocity. If we on the earth, who move through space along with the sun, look at either the average radial velocities or the average proper motions of nearby stars in different parts of the sky, we can see a reflection of the solar motion. For example, with the proper motions we will see a net motion of the stars from one point in the sky directed toward the opposite point in the sky. As shown in Figure 7–13, the effect will be most pronounced at 90° in the sky from these two points. An example of this is the two views one perceives while riding in an automobile. While looking out a side window, one sees rapid motion across the field of view; however, while looking out a front window, one sees very little movement ahead. Of course, both are simple perspective effects. By tracing proper motions back in the sky to the point where they would intersect, we can determine in which direction the sun is moving. This point, called the *apex* of the solar motion, lies slightly to the south of the bright star Vega. The sun's speed can be found by analyzing the radial velocities of the nearby stars.

Once the solar motion with respect to nearby stars of a given type has been found, we can subtract it from the motions of those stars. For instance, if we look at a group of stars in one region of the sky, the effect of the solar motion on their radial velocities will be essentially the same. The average of their radial velocities would be zero if the sun were not moving through space. Instead we find that the average of the radial velocities is the component of the solar motion in that direction. [This will only be zero if the stars lie 90° in the sky from the apex of the solar motion, at points $B$ and $C$ in Figure 7–13(a)]. But we are not interested in the average radial velocity; instead we want the average *size* of the radial velocity, regardless of whether the radial velocity is positive or negative. We also need the average *size* of the proper motion, either the

Figure 7–13. *The sun's motion through space. (a) Reflection of the sun's motion in the motion of nearby stars. We can consider the effect of the solar motion on the observed motions of nearby stars as a motion of those stars in the opposite direction. (b) The solar motion causes nearby stars to appear to be radiating from the* apex, *the direction of the sun's motion, towards the* antapex. *The proper motions of the stars converge towards the antapex.*

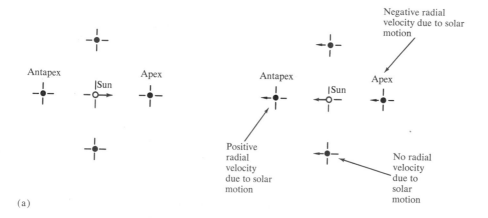

(a)

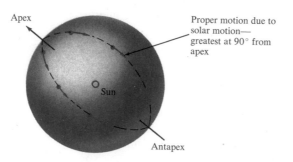

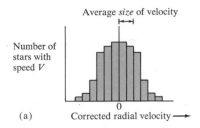

(a)

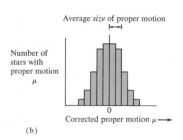

(b)

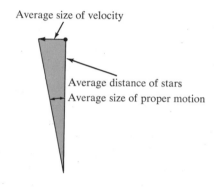

(c)

Figure 7–14. *Statistical parallax. (a) Schematic histogram (distribution) of radial velocities for stars in one part of the sky. The radial velocities have been corrected so the average (not average size) is zero. Thus the effect of the solar motion has been taken out. (b) Schematic histogram of proper motions in one direction on the sky for stars in one part of the sky. As before, the effect of the solar motion has been taken out by correcting the proper motion so the average is zero. (c) Relationship between average size of proper motion, average velocity, and average distance. Note resemblance to Figure 7–3.*

part parallel to the direction of the solar motion in that part of the sky [Figure 7–13(b)] or the part perpendicular to it. Since the proper motion depends on both the tangential velocity and the distance of each star, we have to make a correction to the average proper motion for the different distances of the stars. Once we have the average size of the radial velocity, we assume that the average size of the velocity in any other direction is the same (which will be true if the stars' motions are actually random). The average sizes can be found from the distributions of radial velocities and proper motions, as indicated in Figure 7–14(a) and 7–14(b). Once these are found, we use them as shown in Figure 7–14(c) to find the average distance.

### Dynamical Parallax

Another indirect method is the *dynamical parallax*. It uses a relationship between the mass of a star and its absolute magnitude. If we have a visual binary, we know that Kepler's third law will apply:

$$m_1 + m_2 = \frac{a^3}{P^2} \qquad (7\text{–}12)$$

where $m_1$ and $m_2$ are the masses of the two stars relative to the sun's mass, $P$ is the period of revolution of the pair in years, and $a$ is the semimajor axis in a.u. We can measure the apparent magnitudes $m_1$, $m_2$ and the apparent size $a'$ arc-sec of the orbit. We assume here that we can correct for the foreshortening caused by the tilt of the binary star orbit to our line of sight. (See Figure 7–15.) To begin, we assume that the sum of the masses is 2, so that

$$a = \sqrt[3]{2 \cdot P^2} \text{ a.u.} \qquad (7\text{–}13)$$

The distance is then given by

$$d = \frac{a}{a'} \text{ parsecs} \qquad (7\text{–}14)$$

(This is basically the same formula as for trigonometric parallax.) From the distance $d$ and $m_1$ and $m_2$ we can find the absolute magnitudes $M_1$ and $M_2$. For reasons we will go into later (Chapter 9), the normal (dwarf) stars like the sun (that is, luminosity class V) obey a certain relationship between such a star's mass and its absolute magnitude. So if we have values for the absolute magnitudes, this relationship gives us new values for the masses—$m_1'$, $m_2'$. These are not the correct values because the distance $d$ was not correct, so we substitute again:

$$a = \sqrt[3]{(m_1' + m_2')P^2} \text{ a.u.} \qquad (7\text{–}15)$$

Thus we find new absolute magnitudes. After two or three tries this procedure usually gives the correct result for $d$, $m_1'$, and $m_2'$.


**Membership in Clusters**


Last, there is a trivial but very important indirect method. If we can find the distance to one star in a star cluster, we have found the distance of the cluster and all its stars are virtually the same distance from us. This trick gives us absolute magnitudes for a large group of stars as long as we know how much interstellar absorption there is between us and the cluster. The only real problem is in making sure that a given star belongs to the cluster and is not merely a foreground or background star. The best way of doing this is to see whether the star's motion (radial velocity or proper motion) is nearly the same as the cluster's.

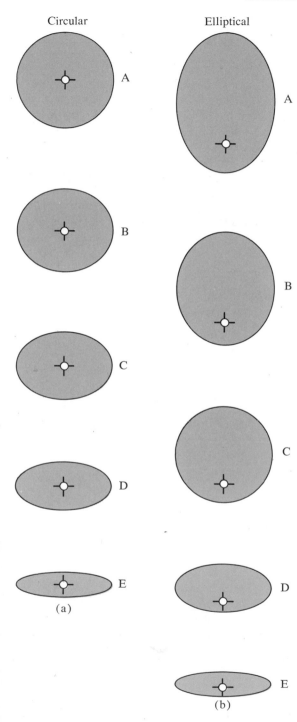

Figure 7–15. *Tilt of the orbit of a binary star. (a) Circular orbit viewed at several different inclinations, from face-on (A) to edge-on (E). The orbit appears more highly eccentric as the tilt increases. Note that the primary star is located at the* center *of the ellipse, not at the focus as it should be for a truly elliptical orbit. (b) An elliptical orbit viewed at several different inclinations, from face-on (A) to edge-on (E). The orbit appears circular at C, but the primary star is not at the center, as it should be for a truly circular orbit. The shapes shown here are only for one orientation of an elliptical orbit. It is the location of the primary that allows us to determine the true shape, orientation, and inclination of the orbit.*

Questions

1. What is the distance in parsecs of a star whose parallax is 0.04 arc-sec?

2. We have found that a star cluster is moving at 20 kilometers per second at a right angle to the line-of-sight to the cluster. Its proper motion has been measured and found to be 0.1 arc-sec per year. What is its distance in parsecs?

3. The Crab Nebula, the shell thrown off by a supernova that exploded in 1054 A.D., is expanding with a velocity of 1,100 kilometers per second (measured from Doppler shift). The proper motion of the shell's edge due to expansion is 0.21 arc-sec per year. How far away is it?

4. What would the apparent magnitude of the sun (absolute magnitude + 4.8) be if it were at a distance of 1,000 parsecs and if there were no interstellar absorption? What would its apparent magnitude be for the same distance if the absorption were 2.5 magnitudes?

5. Using either a table of antilogarithms or Figure 7–9, find the distances of the following stars:
   (a)  $m = +5.0, M = -5.0$, abs $= +1.5$
   (b)  $m = +15.0, M = 0.0$, abs $= +3.0$
   (c)  $m = +8.0, M = +16.5$, abs $= 0.0$

6. How many kilometers are there in a parsec? (The astronomical unit is approximately 150,000,000 kilometers.)

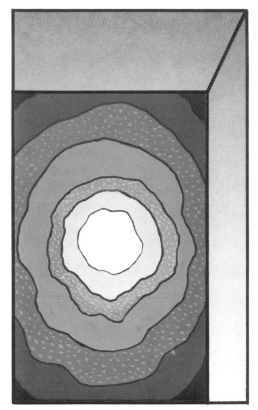

# 8 Finding the Basic Properties of Stars

## Introduction

In Chapter 6 we saw that some of the important observable quantities of a star are its apparent brightness (its *apparent magnitude*), its color, its spectrum, and its membership in a star system. Now we will see how these and other observational data are used to estimate the basic properties.

First, however, we should clarify the meaning of the terms *light curve* and *radial velocity curve*. As mentioned in the preceding chapter, we can plot observations of a star's apparent magnitude against the time of each observation (shown in Figure 7–11). If the star is a variable star and the variation is periodic, we can determine the period and overlap successive cycles. Any such plot is called a *light curve*. Similarly, with any star whose radial velocity (motion towards or away from us as measured from the Doppler effect) varies, we can plot observations of the radial velocity with time, as shown in Figure 8–1. Examples of stars whose radial velocity varies are spectroscopic binaries and pulsating variable stars (whose surfaces move in and out). As before, if the variation is periodic, we can overlap successive cycles. Such plots are called *radial-velocity curves*.

## Luminosities

We might suppose that once the distance to a star has been found, measurement of its apparent magnitude would immediately give its light output or luminosity from Equation (7–11):

$$m - M = 5 \log r - 5 + \text{abs}$$

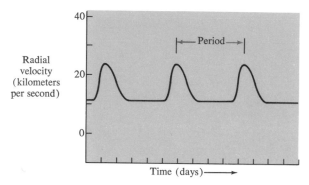

Figure 8–1. *Radial-velocity curve (schematic) for a single-line spectroscopic binary.*

where $m$ is the apparent magnitude, $M$ the absolute magnitude, and where abs is the amount of absorption in magnitudes between the star and earth. (As pointed out in the preceding chapter, the absorption is usually difficult to determine in practice.) But we must remember that apparent magnitude usually refers to just *one* small region of the spectrum; we are interested in finding the *total* output of electromagnetic radiation at all wavelengths. The stars, like blackbodies (Chapter 3), radiate at all wavelengths.

. So we add to the absolute magnitude a term that allows for radiation at nonvisible wavelengths, called the *bolometric correction,* abbreviated B.C. (The term "bolometric" refers to the *bolometer,* a device for measuring radiant energy at all wavelengths.) We then have the *bolometric absolute magnitude:*

$$M_{bol} = M + \text{B.C.} \tag{8–1}$$

Since the correction is for *extra* radiation, making the star *brighter* overall, the bolometric magnitude will be less than the absolute magnitude. This means that, as the equation is written above, the bolometric correction will be *negative*. We recall that for blackbodies the proportion of energy radiated at short wavelengths increases with increasing temperature. This means that stars hotter than the sun will radiate a greater proportion of their energy at short (that is, ultraviolet) wavelengths than the sun radiates at short wavelengths. Then for surface temperatures greater than the sun's, the bolometric correction becomes more negative as the temperature increases. The proportion radiated at long (that is, infrared) wavelengths increases as the surface temperature decreases, so at low surface temperatures the bolometric correction is negative also. For stars slightly hotter than the sun (F0 stars, with $T = 6500°$), the

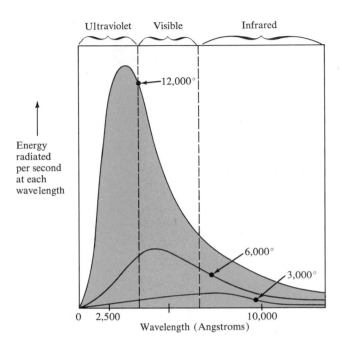

Figure 8–2. *Illustration of the need for a correction for radiation emitted at wavelengths outside the visual region of the spectrum. As can be seen, all stars—including the sun—emit some radiation at all wavelengths. However, not all stars emit the same* fraction *of their total energy output in the visible part of the electromagnetic spectrum. Stars with low surface temperatures (e.g., T = 3000°K) emit a larger proportion of their radiation at infrared wavelengths than stars like the sun (T = 6000°K). Stars with high surface temperatures (T = 12,000°K) emit a larger proportion of their radiation at ultraviolet wavelengths than stars like the sun. The correction is referred to as the* bolometric correction *(see text).*

bolometric correction is zero. The bolometric corrections for stars and for blackbodies are shown in Figure 8–3.

These bolometric corrections are somewhat difficult to calculate for stars. In the days before space probes, measurements of radiation from stars at infrared and ultraviolet wavelengths were severely handi-

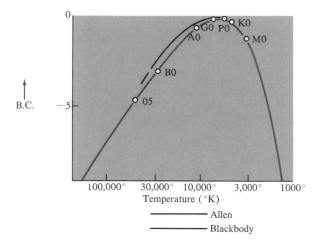

Figure 8–3. *Bolometric correction as a function of temperature. The blue curve is for stars, while the black curve is an approximation for a blackbody. Spectral types are indicated; actually, however, the correction depends on the luminosity class as well as spectral type.*

capped by absorption in the earth's atmosphere. Except for the visible region and a few small bands at longer wavelengths, most of the radiation is absorbed. This is fortunate for us, because it means we are protected from the sun's strong ultraviolet radiation. But in the past it also forced us to rely on theoretical calculations of the bolometric correction. Now we have satellites outside the earth's atmosphere with detectors that can measure the ultraviolet and infrared radiation from stars. (In the recent past rockets and balloons have also been used for this purpose.)

## Masses

### Masses from Kepler's Third Law

In a binary system, the masses of stars are determined by using Kepler's third law (see Chapter 7):

$$m_1 + m_2 = \frac{a^3}{P^2}$$

where as before $m_1$ and $m_2$ are the masses of the respective stars, $a$ is the semimajor axis of their orbit around each other, and $P$ is the period of revolution. For the above equation to be correct as it is written, we must always take care to use the correct units. With $a$ given in astronomical units and $P$ given in years, the masses will be in solar masses, that is, relative to the sun's mass. Notice that the equation only gives the *sum* of the masses, not the mass of each star. If we can determine the ratio of the two masses, $q$, and the sum of the masses as well, we have two equations in two unknowns. It is easy to solve for the mass of each star, as follows:

$$q = \frac{m_1}{m_2} \qquad (8\text{-}2)$$

Cross-multiplying,

$$m_1 = qm_2 \qquad (8\text{-}3)$$

so

$$m_1 + m_2 = qm_2 + m_2 = m_2(q + 1) = \frac{a^3}{P^2} \qquad (8\text{-}4)$$

Then

$$m_2 = \frac{a^3}{P^2} \times \frac{1}{q + 1} \qquad (8\text{-}5)$$

And finally this means that

$$m_1 = qm_2 = \frac{a^3}{P^2} \times \frac{q}{q + 1} \qquad (8\text{-}6)$$

### Semimajor Axis of the Orbit

The first problem is to find the actual values of $a$ and $P$. If the system is an eclipsing binary or a spectroscopic binary, the period $P$ can easily be found from the light curve in the first instance (eclipsing binary) or the radial-velocity curve in the second (spectroscopic binary). If we are dealing with a visual binary, the period is likely to be fairly large (on the order of a century), but in most cases it is still possible to estimate it with high accuracy.

Finding the length of the semimajor axis $a$ can present difficulties. With a visual binary, we can determine the tilt of the orbital plane to our line of sight and correct for the projection effects. (Knowledge of the geometric properties of the ellipse enables us to do this. See Figure 7–15.) Once the angular extent of the orbit (in arc-seconds) has been found and corrected for projection, we can use the distance of the system to convert that angle into astronomical units:

$$a = a' \times r \text{ a.u.} \tag{8-7}$$

with $a'$ in arc-sec and $r$ in parsecs. If the binary is eclipsing, we can determine the tilt from the shape of the light curve. In that case, however, we need the radial-velocity curve to find the projected size of the semimajor axis. Thus the system must be both an eclipsing binary and a spectroscopic binary.

This process can be easily illustrated for the simplest case—that of circular orbits. The radial-velocity curve for this case is shown in Figure 8–4. The horizontal line represents the velocity of the system's center-of-mass, or the motion of the system as a whole. The vertical line represents the radial velocities of the two stars relative to each other. If this velocity is $V$, the projected orbital size $a$ is given by

$$2\pi a = V \times P \tag{8-8}$$

or

$$a = \frac{V \times P}{2\pi} \tag{8-9}$$

(We would have to express $V$ in a.u. per year.) This value can be corrected for the projection effect very easily. If the orbits are not circular the analysis becomes more complicated, but it can still be performed using the radial-velocity curve.

### Determination of the Mass Ratio

To find the mass ratio $q$, we must find either the distance of each star from the center-of-mass or the velocity of each star with respect to the center-of-mass. If the system is a visual binary, we locate the center-of-mass with respect to the background or reference stars. We can see how this is done if the system has no proper motion; that is, if as a whole its position with respect to the reference stars does not change. Figure 8–5 shows a plot of the positions of the two stars at several different times for a hypothetical visual binary. The center-of-mass must always lie on a line joining the two stars, and this fact is not changed by projection effects. So if we draw lines connecting the positions of the two stars at each instant, as shown in the diagram, these lines will cross at the center-of-mass. (If the system has some proper motion, we must first correct for it before drawing the lines.) The distances of the stars from the

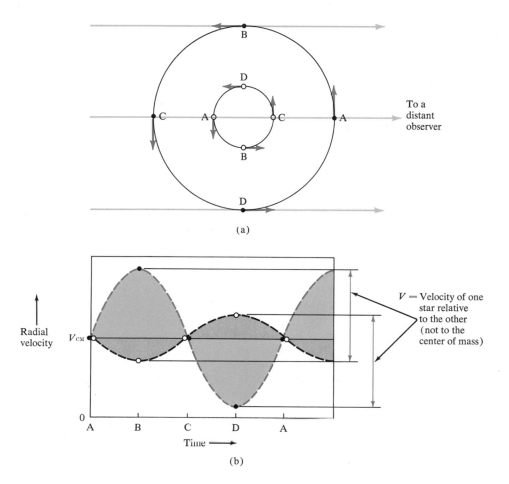

(a)

(b)

Figure 8–4. *(a) Binary with circular orbits around the center-of-mass, the simplest case. The arrows represent the velocities of the two stars with respect to the center-of-mass. The stars orbit around the center-of-mass as it moves through space in a straight line at a constant speed. (b) Radial-velocity curve for the system as a double-line spectroscopic binary. The solid circles represent values for the less massive star, and the open circles represent values for the more massive star (the one nearer the center-of-mass). The times of the observations correspond to the letters in Figure 8–4(a). The velocity of the one star with respect to the other, the relative velocity V, is indicated at right.*

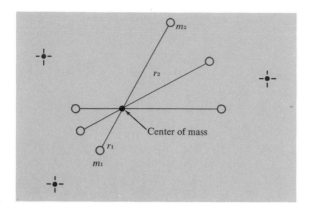

Figure 8–5. *The mass ratio for a visual binary. The ratio of the distances of the stars from the center of mass, which is located at the intersection of the lines, gives the ratio of the masses. A similar, though more involved, procedure can be used in the more typical situation when the center-of-mass moves on the sky. (See also Figure 6–11.)*

center-of-mass at any instant are in inverse proportion to the ratio of their masses, since this is how the center-of-mass is defined. (It can be thought of as a balance point for the system.) We then have

$$q = \frac{m_1}{m_2} = \frac{r_2}{r_1} \tag{8-10}$$

The terms $r_1$ and $r_2$ are the respective distances from the center-of-mass.

If the system is a double-line spectroscopic binary, we can find the mass ratio from the amount of variation in the velocities of the two stars, $V_1$ and $V_2$, measured from the radial-velocity curve as shown in Figure 8–6. Then

$$q = \frac{m_1}{m_2} = \frac{V_2}{V_1} \tag{8-11}$$

## Radii

### Angular Diameters of Stars

The actual sizes of stars (for example, in miles) can be determined from their apparent sizes in the sky (angular diameter) and their distance if we can measure their apparent sizes. Apart from the sun

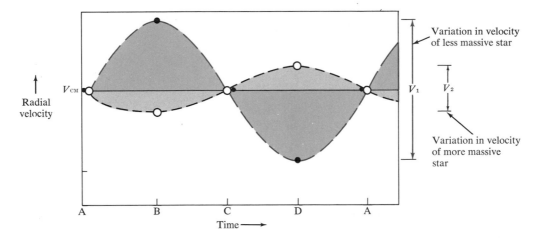

Figure 8–6. *The mass ratio for a spectroscopic binary. The radial-velocity curve from Figure 8–4(b) is shown, together with the variations in velocity for the two stars $V_1$ and $V_2$ as indicated at right. As before, the solid circles represent values for the less massive star, and the open circles represent values for the more massive star.*

and a very few other stars, however, we cannot do this. The sun is of course near enough that it appears relatively large in the sky ($1/2°$). In the other cases where we can measure the apparent sizes, the measurements are very difficult. One way is to measure the light from the star as it is *occulted,* or hidden, by the moon. The rate at which the light decreases (or increases, when the star reappears from behind the moon) gives information on the star's apparent size, since we know how fast the moon moves against the star background. The occultation method only works for bright stars near the ecliptic in the sky. Another way of measuring apparent sizes involves the use of *interferometry.* In this method, signals from two or more widely separated mirrors are mixed, either optically or electronically. The interferometer can only be used on stars with relatively large apparent sizes, of the order of 0.001 arc-sec or greater. This may not seem very large (the apparent size of a dime at a distance of approximately 900 kilometers), but it is much greater than the typical apparent size of a star.

### Radii of Members of Eclipsing-Spectroscopic Binaries

Major sources of information on stellar radii are the eclipsing-spectroscopic binaries. Analysis of the light curve gives us the sizes of the two

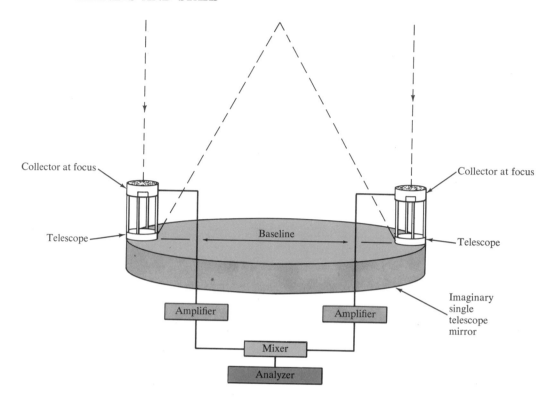

Figure 8–7. *The basic principle of the inter-ferometer. Signals from the two telescopes are amplified and mixed electronically (or instead simply run together optically) and the resulting interference signal analyzed. The interferometer has imaging properties similar to those of an imaginary single reflecting telescope with its mirror covered except for two small patches, one on each side. High resolution, like that of the imaginary large telescope, is obtained in a direction on the sky along the baseline; in the direction per-pendicular to the baseline the resolution is only that of one of the small telescopes.*

stars relative to their average separation. We can see this fairly easily for an eclipsing binary system with a circular orbit seen edge on. Figure 8–8 shows the light curve for such a hypothetical system. The time taken by the *eclipsing* star in moving from *a* to *b* is the same fraction of the period of revolution as the diameter (of the star) is of the circumference

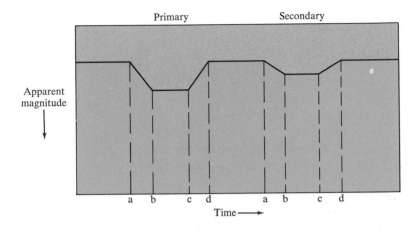

Figure 8–8. *Idealized light curve for an eclipsing binary. The times a,b,c,d, are the times of the* contacts *of the eclipse, with primary eclipse the deeper (lower brightness). First contact (a) is the beginning of the first partial phase of the eclipse, second contact (b) is the beginning of the total phase, third contact (c) is the end of the total phase, and fourth contact (d) is the end of the second partial phase.*

of the orbit, as shown in Figure 8–9. Mathematically, we express this as

$$\frac{t_{ab}}{P} = \frac{2R_1}{2\pi r} \qquad\qquad \textbf{(8–12)}$$

where $R_1$ is the radius of the *eclipsing* star and $r$ is the radius of the orbit. By similar reasoning, the time for the eclipsing star to move from $a$ to $c$ is the same fraction of the period of revolution as the *eclipsed* star's diameter is of the circumference of the orbit, as shown in the Figure 8–9. Then

$$\frac{t_{ac}}{P} = \frac{2R_2}{2\pi r} \qquad\qquad \textbf{(8–13)}$$

To find $R_1$ and $R_2$, we need to know $r$. The spectroscopic observations give us the orbital speed of the one star with respect to the other, $V$ (shown in Figure 8–4). For a circular orbit, we have this equation:

$$2\pi r = V \times P$$

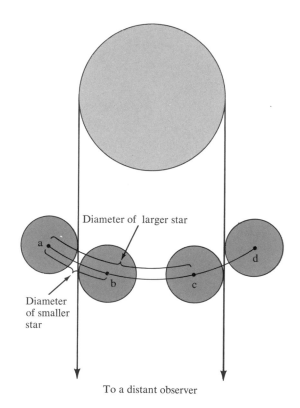

Figure 8–9. *Timing of contacts for the special case of a central eclipse (one star passing directly in front of the other) and circular orbits. The points a,b,c,d correspond to the times indicated on the light curve. In the time the smaller star moves in orbit around the larger from a to b, it moves along a distance equal to its own diameter. In the time the smaller star moves from a to c, it moves along a distance equal to the larger star's diameter. (Follow the edge of the smaller star.) Note that if the observer were in the opposite direction, with the larger star eclipsing the smaller, the distances ab and ac would be the same.*

### Complications with Eclipsing Binaries

There are, however, a number of complications to this problem. Usually the orbits are elliptical instead of circular, and the orbit is seldom seen exactly edge on. (It must be nearly so or there would be no eclipses.) Also, we have assumed that both stars are spherically symmetric (always presenting a circular cross-section) and uniformly bright all over their surface. At least one, and often both, of these conditions are not satisfied.

For one thing, when two stars are very close together they exert extremely strong tidal forces on each other. Like the earth's oceans under the influence of the moon's tidal force, the gaseous material of which the stars are made is stretched out into an elongated shape, somewhat resembling a "blimp" or a fat cigar. When these *tidally-distorted* stars are at or near an eclipse, we are looking down their long axes. At that time, they would appear circular if we could see them. Later, after a quarter-period, we would see them from the side. At that time, they would be presenting more surface area to us, so that overall they would appear brighter. As a result, the light curve appears "bowed up" between eclipses. Also, all stars would appear slightly darker near the edge (or

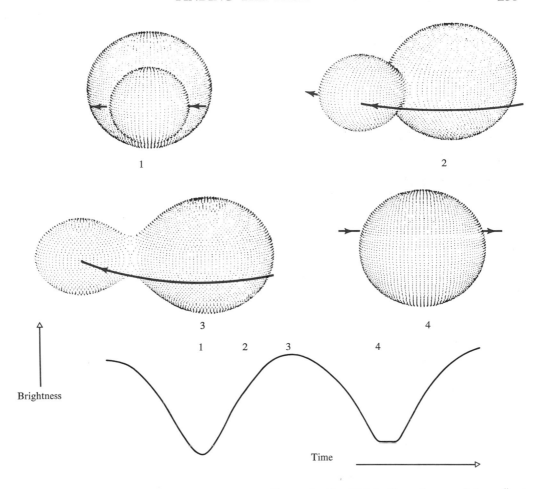

Figure 8–10. *Tidal distortion and its effect on the light curve.*

*limb*) of their disk than at the center, if we could see them as disks as we see the sun. Photographs of the sun show this effect, called *limb-darkening,* very clearly. We can see deeper into the star at the center than at the limb because light from the latter travels farther through the star's atmosphere, even though it starts at a higher level. This is demonstrated in Figure 8–11. Since the atmosphere is cooler at the higher level, the star appears less bright at the limb. The limb-darkening makes the eclipses less abrupt, so that the light curve is rounded off, as Figure 8–'2 shows.

Another complication is the *reflection effect.* When a hot star is paired with a cool one, and the two are fairly close together, the hot star's radiation will heat the nearest part of the cool one's surface. The

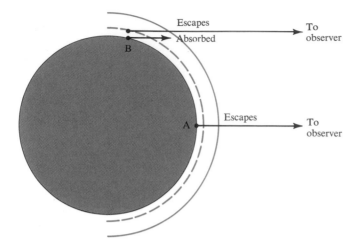

Figure 8–11. *The cause of limb-darkening. Imagine that you are looking at the star from the right. The light from point A, at the center of the star's disk, can penetrate the star's atmosphere. The light from point B cannot reach the observer because it must travel a greater distance through the star's atmosphere. At that point, light from higher levels can reach the observer. At these higher levels the temperature is lower, so at the edge of the disk—the* limb*—the star does not appear as bright.*

result is a "hot spot" on the cool star which is brighter than the rest of the star. When the hot star passes behind the cool one, the "hot spot" will be invisible because it is facing away from us. When the hot star passes in front of the cool one, the "hot spot" will be visible before and after the eclipse and invisible during the eclipse. Since the area eclipsed is the same in both cases if the orbit is circular, the *deeper* minimum in the light curve *(primary minimum)* is the eclipse of the *hotter* star, which is brighter per unit area. Because of the reflection effect, the light curve is tilted upwards on either side of the secondary minimum (Figure 8–13).

Each of these effects causes its own distinctive change in the light curve, and a mathematical analysis can unravel the contribution of each. As this simplified description shows, though, it is a tricky problem to extract radius values.

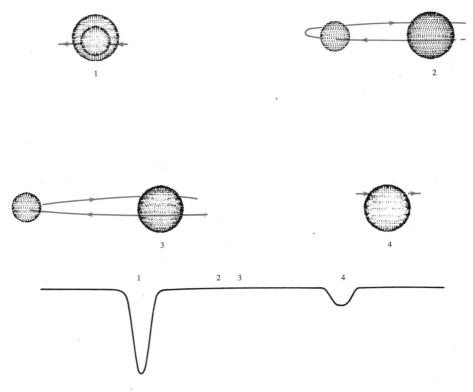

Figure 8–12. *Effect of limb-darkening on the light curve of an eclipsing binary.*

## Temperatures

### Estimates from Radiation Laws

The main difficulty in determining the surface temperatures of stars arises because the temperature—the average energy of motion or kinetic energy of the atoms—is different from one level in the star's atmosphere to another. The absorbing power of the gases varies with the wavelength of the light, so we can see farther into the star at some wavelengths than others. The starlight is thus a mixture of light from different levels, each at a different temperature, rather than light from a well-defined surface at a given temperature. Still, as a first approximation, we can compare the star's spectrum to that for an ideal

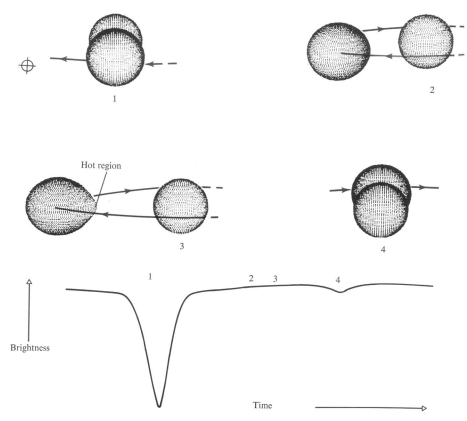

Figure 8–13. *"Reflection effect" on the light curve of an eclipsing binary.*

radiating body. The temperature of the latter can be found by comparing the amount of energy radiated at one wavelength to the amount radiated at another wavelength. Since this is a measurement of a color, we refer to the temperature determined in this way as a *color temperature.* Or we can find the temperature by measuring the total energy radiated by each unit of area per second. If the total energy is $L$, the temperature is $T$, and the radius of the (spherical) object is $R$, we have

$$L = \sigma T^4 \times 4\pi R^2 \qquad \text{(8–14)}$$

where $\sigma$ is a constant of proportionality. If we solve this equation for $T$, we find

$$T = \sqrt[4]{\frac{L}{\sigma \times 4\pi R^2}} \qquad \text{(8–15)}$$

For a star, we use the luminosity for $L$ and the radius of the visible surface, or photosphere, of the star for $R$. The temperature calculated this way is called an *effective temperature*. Unlike the color temperature, which will in general depend on our choice of the two wavelength intervals, the effective temperature does not depend on wavelength. For this reason, it is a more meaningful quantity physically. Accordingly, we try to relate the other indicators of surface temperature, such as color and spectral type, to the effective temperature.

### Estimates from Model Stellar Atmospheres

As we indicated in Chapter 6, the spectral type depends on surface temperature. This is because the strengths of spectral lines of a given element depend on the temperature as well as on the amount of that element present in the star's atmosphere. As we will see in the next section, the various elements occur in nearly the same proportions almost everywhere. This means most of the differences in strength that we see are caused by temperature differences. The strength of a given spectral line depends on the temperature, which determines the fraction of atoms of that element in the particular ionization and excitation stage causing the line (see Chapter 3). The behavior of some of the more common lines is shown in Figure 6–6. Now we can understand the spectral-classification sequence. The strengths of various spectral lines allow us to estimate the surface temperature.

To actually do this with precision, however, it is necessary to calculate a "model stellar atmosphere." The model is actually a set of values of pressure, density, and temperature at numerous different depths in the star's atmosphere. These numbers are solutions of the complicated mathematical equations that describe the transfer of energy and the balance of forces in the gas of the star's outer layers. Among other things, we can use the model to predict the strengths and shapes of the various spectral lines. Each model is characterized by a value of effective temperature, surface gravity, and chemical composition. It becomes simply a matter of finding the model that best matches the observations. Ultimately, from the observations and from theoretical models, we find relationships among color, spectral type, and effective temperature. For example, the values for main-sequence stars are shown in Table 8–1.

## Chemical Composition

### Spectral Line Profiles

As we have seen previously, the strengths of the various spectral lines of a given element depend on the excitation and ionization

## TABLE 8–1

*Temperatures for Main-Sequence Stars*

| Spectral type | Color | Temperature (°K) |
|:---:|:---|:---|
| 05 | blue | 35,000 |
| B0 | blue-white | 21,000 |
| A0 | white | 9,700 |
| F0 | yellow-white | 7,200 |
| G0 | yellow | 6,000 |
| K0 | orange | 4,700 |
| M0 | red | 3,300 |

SOURCE: C. W. Allen, *Astrophysical Quantities,* 2d ed. (New York: Oxford University Press, 1963).

conditions in the star's atmosphere. These in turn depend on the temperature and also on the density and pressure. But the strengths must also depend on the abundance of the particular element. (Note that the term "abundance" does not imply that there are large amounts of the element present, at is does in common parlance.) If there is none of that element present, none of its spectral lines will be present.

Before discussing the relationship between the relative abundances and the strengths of spectral lines, we need to discuss the *strength* of a line further. The shape of a spectral line (the *line profile*) depends on a

Figure 8–14. *(a) Energy radiated as a function of wavelength (schematic). The thin line indicates the continuum, i.e., the spectrum if no lines were present. (b) Energy radiated as a percentage of the continuum at each wavelength. Note that the two most prominent absorption lines, which in (a) seem about the same strength, have different strengths in (b).*

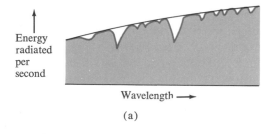

(a)

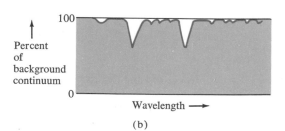

(b)

number of factors. According to the quantum theory, every spectral line has a certain natural shape, apart from all other influences. One of the most important of these "other influences" is that of atoms' random thermal motions. These motions cause Doppler shifts in the wavelengths of photons emitted by the atoms. This *thermal broadening* will become greater as the temperature is increased. Pressure and magnetic and electric fields also broaden the lines. If the star is rotating and we are not looking along its axis of rotation (or "pole-on"), light from different regions of its surface will be Doppler-shifted by the rotation. As a result, the spectral lines take on a "dished-out" or smeared appearance, with the amount of smearing dependent upon the projection of the rotational velocity at the star's equator. Since these various effects act differently on the shapes of the spectral lines, we want some indicator of a line's strength that does not depend on the exact shape. The most satisfactory such quantity is the *equivalent width* of the line. As the name implies, it is the width of a strip of spectrum that has the same area as the portion of spectrum taken up by the line itself (see Figure 8–15).

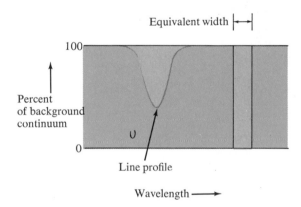

Figure 8–15. *Equivalent width of an absorption line. The rectangular strip has the same area as the shaded portion of the line profile. The width of the rectangular strip is the equivalent width.*

The equivalent width is related to the number of atoms. The relation is presented as a *curve of growth* (Figure 8–16). For weak lines, the equivalent width increases in proportion to the number of atoms. As the line becomes *saturated,* so that essentially all the light is blocked out at the center of the line, the equivalent width changes very slowly. However, as the number of atoms increases still more, the *wings* of the line (the outlying portions) become more pronounced, and the equivalent width increases in proportion to the square root of the number of atoms.

The abundances of the elements can be found by comparing curves of growth for different elements. A more accurate procedure, however, involves the use of model stellar atmospheres, as described in the preceding section. The models are used to predict the shapes and strengths of various spectral lines for comparison with observations.

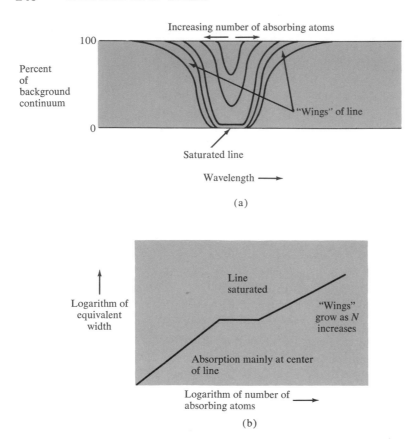

Figure 8–16. *(a) Shape of an absorption line as the number of absorbing atoms increases. The center of the line (the "core") gets deeper until it reaches zero, when the line becomes saturated. Then the line gets wider, with the "wings" of the line spreading out. (b) Increase of equivalent width as the number of absorbing atoms increases, called the* curve of growth *for the line.*

## Results of Studies of Chemical Composition

Generally, the studies of the abundances of the various elements in stars show that they are roughly the same in all stars. With a few exceptions, most stars consist primarily of hydrogen, a sizeable amount of helium, and traces of the other elements (especially carbon, nitrogen, oxygen) in varying amounts. The variations in abundances may be due either to differences in the initial composition (for example, some stars may have started with more helium than others) or to chemical evolution as the various nuclear reactions proceed in the star. The present belief is that all of the heavy elements (that is, those other than hydrogen and helium) have been produced by the nuclear reactions in stars. The buildup of heavier elements in evolving stars was followed by a large portion of that material being dispersed into interstellar space, as for example in supernova explosions. Later generations of stars would then start life with greater abundances of those elements than the preceding generation. We will return to this idea of "enrichment" later.

## Questions

1. The absolute magnitude of a star is $-1.0$, and its bolometric correction is $-3.0$. What is the star's absolute bolometric magnitude? If the sun's absolute bolometric magnitude is $+4.7$, how many times brighter is the star than the sun? (See Chapter 4 on magnitudes.)

2. For a binary star, the semimajor axis $a$ (corrected for projection) is 4 a.u., and the period $P$ is 3 years.
   (a) What is the sum of the masses of the two stars (in units of the sun's mass)?
   (b) If the mass ratio $q$ is 2.5, what are the masses $m_1$ and $m_2$ of the two stars?

3. A star whose distance is 20 parsecs is found from interferometer measurements to have an angular diameter of 0.01 arc-sec. What is the actual diameter of the star? (Hint: Figure out how large 1 a.u. would be at 20 parsecs, and use the conversion from a.u. to kilometers from Problem 6, Chapter 6.)

4. Describe the causes and effects of three complications encountered in studying eclipsing binaries.

5. Star A has a surface temperature twice that of Star B and a radius one-fourth as large. How much brighter (or fainter) will Star A be than Star B? Which will be *intrinsically* redder (referring to the star's true color)?

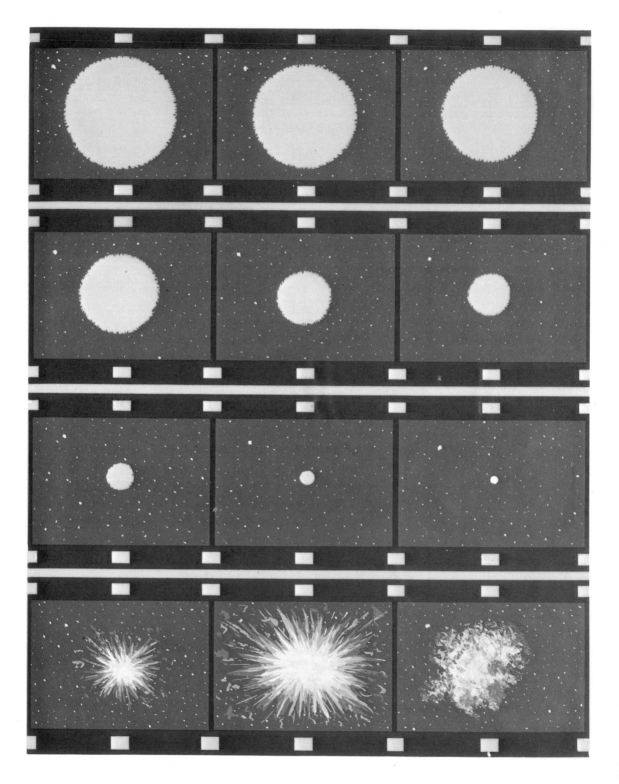

# Stellar Evolution

## Introduction

It is an interesting fact, established by the record of fossils, that the sun's luminosity has remained essentially constant for hundreds of millions of years. Together with the fact that most stars are observed not to vary significantly in brightness or spectrum over decades, this indicates that stars generally evolve very slowly. There are exceptions, namely the variable stars mentioned before, but the typical star shows no measurable change. On the other hand, we do see a bewildering variety of stars around us, including a few that do show spectacular changes. These facts suggest that evolution of the stars does take place, even if it happens to be too slow for us to see. It is as if we have been given a huge jumble of "snapshots" of stars (all to the same scale and exposure) and asked to arrange these into motion pictures showing the evolution of each kind of star. This may seem a hopeless task, but two factors make it possible for us to assemble reasonably complete motion pictures. First, the "snapshots" include some group portraits—stars belonging to a given cluster. We can reasonably assume that all those stars were formed from the same material (though we have not said what that material is) and at approximately the same time. Presumably, then, all those stars had the same chemical composition at the beginning and are all the same age. So, if we can determine the age of one star in the cluster, we then know the ages of all stars in that cluster. Second, we can predict with certainty some of the phases of stellar evolution, thanks to the advances in physics during the present century and to the development of high-speed electronic computers. These predictions must of course be carefully compared with and related to the observational data in order to be sure that the calculations have been done correctly. We will go into this aspect in more detail later.

## The Hertzsprung-Russell Diagram

### Definition and Description

The group portraits referred to above are not really "snapshots," but they amount to much the same thing. Suppose we plot the luminosity of each star of a cluster against its effective temperature on a graph. The range of values is quite large, so we use a geometric or *logarithmic* scale instead of the usual linear scale. Each of the large scale divisions on the horizontal axis represents an increase of a factor of ten; those on the vertical axis represent increases of a factor of 100. Such a graph is called a *Hertzsprung-Russell diagram,** usually abbreviated by astronomers as *H-R diagram*. The different regions of the diagram have names that are related to the luminosity classes of Chapter 6 and to the colors of the stars in those regions. (Remember that color, spectral type, and effective temperature are all closely related.) For example, the region at the upper right is referred to as the *red-giant* region and above it is the region of *red supergiants*. Ordinary dwarf stars fall along the *main sequence,* a curve running diagonally from upper left to lower right. Another region of interest in the H-R diagram is the *white-dwarf* region at the lower left. These main regions are indicated in the H-R diagram in Figure 9–1. There are other features in the H-R diagram that have more specialized names, but we will not discuss them here.

The variety of H-R diagrams for star clusters is great. With some, most of the stars lie along a well-defined main sequence. In others, the main sequence is curved to the right at its upper end, and a number of the cluster's stars lie in the red-giant region. For still others, there is only a short main sequence at the low-temperature end, with very many of the stars in the red-giant region. This last type of H-R diagram is the one that the *globular* clusters have, while *galactic* clusters have the first two kinds of H-R diagram. These differences could be due either to differences in age or to differences in initial chemical composition. (Remember, these are the two factors that are presumably the same for all the stars in a given cluster.)

### Age Estimates for Star Clusters

We can outline very simply the essentials of determining the ages of stars in a cluster. The energy radiated by the stars, as we will see in the next section, is supplied by the conversion of matter into energy. The

---

* Named for Danish astronomer Ejnar Hertzprung and the American Henry Norris Russell, who devised it independently.

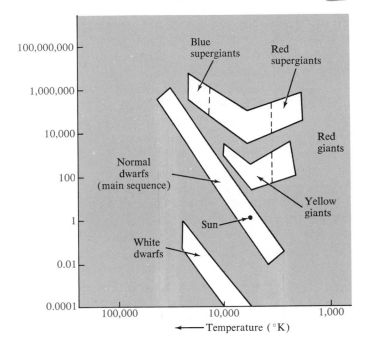

Figure 9–1. *Important regions of the Hertz-sprung-Russell (H-R) diagram. Note the logarithmic scales of luminosity and temperature. The regions shown correspond roughly to the luminosity classes of Chapter 6. The position of the sun is also indicated.*

amount of energy depends on the amount of "fuel," or material available for conversion. Thus,

$$E = c_1 m \qquad (9\text{--}1)$$

where $E$ is the star's total supply of energy, $m$ is the star's mass, and $c_1$ is a constant of proportionality that we assume is the same for all stars. The rate at which stars radiate away this energy—their luminosity—has been shown to be related to the masses of the stars. The *mass-luminosity relation* (derived theoretically by Sir Arthur S. Eddington in 1924, and confirmed by observations) says*

$$L = c_2 m^4 \qquad (9\text{--}2)$$

---

* Actually, the form of the relation is more like $L = c_3 m^{3.5}$ for the least massive stars. This is because they do not exactly satisfy the assumptions Eddington made in deriving the formula.

Figure 9–2. *Hertzsprung-Russell diagram for a galactic cluster (schematic). Each point represents the luminosity and temperature values for an individual star in the cluster. Note several features: (1) the larger number of points on the lower end of the main sequence than the upper; (2) the presence of some highly-evolved stars—red giants and white dwarfs; (3) the gap between the bend at the upper end of the main sequence and the red giants, called the* Hertzsprung *gap after its discoverer and discussed later in the text.*

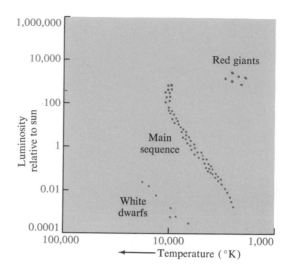

where $L$ is the luminosity, $m$ is the star's mass as before, and $c_2$ is another constant. This mass-luminosity relation is shown in Figure 9–3. The *lifetime* of a star on the main sequence, the time it takes to use up its available energy, is given by

$$t = \frac{E}{L} = \frac{c_1 m}{c_2 m^4} = \frac{c_1}{c_2} \times \frac{1}{m^3} \qquad (9\text{–}3)$$

Equation (9–3) tells us that the stars with the largest masses will use up their "fuel" most quickly. For example, a star twice as massive as the sun will have a main-sequence lifetime of $1/2^3$ or $1/8$ that of the sun. Or, if the star is half as massive as the sun, according to Equation (9–3) it will have a lifetime on the main sequence of $1/(1/2)^3 = 1/(1/8)$ or 8 times that of the sun. By the mass-luminosity relation, the most massive stars are the most luminous. The most luminous stars along the main se-

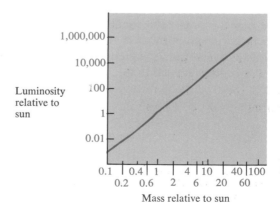

Figure 9–3. *Mass-luminosity relation for normal stars (i.e., those on the main sequence).*

quence are the O- and B-type stars. They will be the first to evolve away from the main sequence, followed later by A- and F-type stars.

With this information, we are now in a position to arrange cluster H-R diagrams in order of age. The age of a cluster is approximately the age of the brightest stars still on the main sequence. The location of this upper end of the main sequence, above which the brighter stars have turned away from the main sequence, is called the *turnoff point*. The farther down the main sequence the turnoff point is, the older the cluster. HR- diagrams for a number of the more prominent galactic clusters are shown superimposed in Figure 9–4. (In order for the diagram to be clear, we have replaced the actual plotted points by shaded regions that indicate where the points fall.) The differences are understandable in terms of the clusters' different ages. When we compare the oldest galactic clusters with globular clusters, we see differences in their H-R diagrams that must not be purely age differences (the turnoff points are almost the same) but, instead, are partly due to differences in chemical composition.

## Nuclear Energy and Stellar Interiors

### Energy Sources for Stars

One of the most significant advances of the present century in physics led to the discovery of the main source of the stars' energy. Ordinary combustion (such as the burning of gasoline) or other chemical reactions are clearly inadequate to supply the sun's energy (to take a typical example) because of the tremendous amount of "fuel" required. Natural radioactivity is also insufficient. For a while the conversion of gravitational potential energy into thermal energy (heat) and radiation was seriously considered as a possibility. If a star is fairly massive and collapses to a very small size, great amounts of energy can indeed be produced. (This process can be likened to the kinetic energy gained by an object, for example, a rock, falling in the earth's gravitational field. When the rock strikes the ground, the kinetic energy is converted into other forms of energy such as heat or mechanical energy. A similar thing happens when the layers of a star fall in its gravitational field.) However, this process (referred to as the *Kelvin-Helmholtz process,* after the physicists who proposed it) is inadequate also. The sun has not shrunk appreciably over geologic time, according to the fossil record, whereas it should have according to the Kelvin-Helmholtz theory. (This does not mean that the Kelvin-Helmholtz process never occurs in stars.)

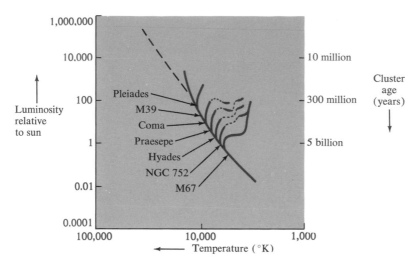

Figure 9–4. *Hertzsprung-Russell diagrams for galactic clusters of various ages superimposed. Bands have been used instead of individual points for the sake of clarity.*

The mystery of the stars' energy was solved in a sense when Einstein derived his famous formula for the equivalence of mass and energy:

$$E = mc^2 \qquad\qquad (9\text{–}4)$$

$E$ stands for energy, $m$ for mass, and $c$ for the speed of light. According to the formula, if only a small fraction of the sun's material could somehow be converted into energy, it would provide more than enough for billions of years.

Methods of doing this in stars were suggested by Hans Bethe and others in the 1930s and 1940s, in what are called *nuclear reactions.* Where chemical reactions involve bonding between atoms of various types, nuclear reactions involve changes in the bonding between subatomic particles in the *nucleus* of an atom or nuclei of several atoms. In particular, we can have *fusion reactions,* where the nuclei of lighter elements are built up to form heavier nuclei; and *fission reactions,* where a heavy nucleus is broken down into lighter nuclei. The hydrogen bomb is an example of the first, while the atomic bomb is an example of the second.

### Nuclear Reactions in Stars

The most important reactions in the interiors of stars are fusion reactions. Three of these are the following:

1.  *Proton-proton chain.* Hydrogen nuclei or *protons* are combined to form a more massive nucleus, of which two are then combined to give a helium nucleus and two protons.* A temperature in excess of 3 million degrees is required.

2.  *Carbon cycle.* In a series of reactions protons are successively added to a carbon nucleus, which finally becomes unstable and splits into a helium nucleus and a carbon nucleus. (The fact that a carbon nucleus results explains why the series is called a cycle.) A temperature in excess of 15 million degrees is required.

3.  *Triple-alpha process.* Three helium nuclei combine to form a carbon nucleus. [A helium nucleus is sometimes referred to as an *alpha particle.* The name alpha particle goes back to an earlier time when the nature of the radiations emitted by radioactive substances was still mysterious. The three types of radiations then known were alpha-rays, beta-rays (now known to be high-speed electrons), and gamma-rays (now known to be very high-energy photons)]. A temperature in excess of 50 million degrees is required for this process.

The first and second reaction series form helium from hydrogen, so they are called *hydrogen-burning* reactions. Note that here the word "burning" does not denote combustion, as in the ordinary sense. The triple-alpha process is a helium-burning reaction. At still higher temperatures we can have carbon- and oxygen-burning reactions, and others as well. With nuclei more massive than that of iron, fusion does not produce energy. Instead the more massive nuclei have to break up (fission) to liberate energy.

As we go to fusion reactions involving nuclei of higher atomic numbers, the temperatures required become higher. This is because of the electric charge of the nucleus. The atomic number, which is the number of protons, gives the number of units of positive electric charge. When two nuclei approach each other, each is repelled by the other's electric charge. The strength of the repulsive force is proportional to the product of the electric charges (in units). The speeds of the nuclei have to be sufficient for them to collide in spite of this force, so if the force is greater the speed must be greater. The temperature is a measure of the mean speed, and at higher temperature the mean speed is greater. Thus higher temperatures are needed to make the heavier nuclei collide.

### Stellar Interiors

The required temperatures are much greater than the surface temperatures of even the hottest stars, so the reactions must be occurring some-

---

* The elementary particles of which atoms are made are described in Chapter 3.

where in the deep interior. To find out just where these are happening, we construct a "model" of the star's interior. (We have to do this because we have no way of actually observing these processes at work inside the star.) The model stellar interior consists of predictions of the temperature, density, pressure, and chemical composition at each of several hundred depths in the star. In effect, we divide the star into a large number of concentric spherical shells. The properties of the material within each shell are assumed to be uniform, while the variations in the several properties from one shell to the next shell are expressed in mathematical formulas called *differential equations*. (These equations contain terms involving the *differences* in certain quantities from one shell to the next shell, as well as the various quantities themselves. This explains why they are called differential equations.) Since the values at all of the shells are interrelated, all the equations must be solved simultaneously. Furthermore, a real star is not comprised of distinct shells, so any solution will necessarily be inexact. Consequently, the electronic computer that does all the calculations is programmed to find a solution of all the equations for all shells that meets some standard of accuracy.

What are these equations? To write them down and discuss them in detail would certainly be out of place in an elementary text. They are very complicated in practice, but they express some very basic physical laws. For example, one equation expresses the balance between the net pressure (difference in pressure from one side of the shell to the other) and the gravitational force on that shell. For the star to be in equilibrium, the net pressure must balance the gravitational force at each point inside the star. The pressure inside each shell depends on the temperature, density, and chemical composition of the material. The relationship among these latter quantities is called the *equation of state* of the gas. The temperature in turn depends on the transport of energy through the shell.

There are three ways in which energy can be transported: *radiation, convection,* and *conduction.* An example of radiative transport is the infrared radiation from a fire, which heats only the side of an object facing the fire. An example of convection is the method of energy transport in the (misnamed) home "radiator," whose metal fins heat the surrounding air. Blobs of heated air then rise to the ceiling, while cooler air moves in to take their place. A circulation system is thus set up in the room. Anyone who has started to pick up a silver spoon that has been sitting in a bowl of hot soup has firsthand experience with a very good example of conduction. The silver (which is a good conductor) carries the heat from the soup to the spoon's handle. If energy is transported by radiation, we must calculate the absorbing power of the material for the different kinds of radiation. Likewise, if transport is by conduction, the conducting properties of the material must be calculated.

In stars, radiative or convective transport usually dominate. If convection is present, with blobs of hot gases rising inside the star and blobs of cooler gas sinking, the material throughout the region where this circulation is occurring gets mixed, much as in a blender. If nuclear reactions have converted sizable amounts of hydrogen into helium in one portion of this convective region, the helium will be spread out into the entire region. We see, then, that we must keep track of the chemical composition of each shell as it is changed by nuclear reactions and/or by mixing.

Another important equation expresses the conservation of energy. The energy leaving a shell must be balanced by the amount of incoming energy plus the energy generated inside the shell (if any) by, for example, nuclear reactions. We also have conservation of matter, since the sum of the masses of the shells must equal the star's total mass. Usually the latter is considered to be constant, since nuclear reactions only convert a tiny fraction of the star's mass into energy. In fact, the mass of the star is the most useful quantity to use in describing the evolution of the different types of stars. The other basic quantities such as luminosity, surface temperature, and radius all change for a given star in the course of its evolution. For a given star, the mass usually does not change. (As we will see shortly, there are circumstances where this is not true.)

## Interstellar Medium, Nebulae, and Star Formation

### Dust and Gas in Interstellar Space

We turn now to the results of the study of stellar evolution. First we consider how the stars are formed. There seems to be ample raw material for the formation of stars lying around even now, namely interstellar dust and gas. This material is generally referred to as the *interstellar medium,* since it is located in the space between the stars. As we have mentioned before, the presence of the dust is indicated by the absorption of starlight and especially by the reddening of starlight, which has been compared to the reddening of the setting sun caused by dust and molecules in the earth's atmosphere. Also, there are dark areas in the sky, such as the "Great Rift" in the constellation Cygnus, which are silhouetted against the background star clouds of the Milky Way (see Figure 9–5). We see *reflection nebulae* surrounding some stars, such as those in the Pleiades. (See color plate 10.) They reflect the starlight so they have nearly the same colors as the stars they surround. Dust reflects light in this manner. The presence of interstellar gas is

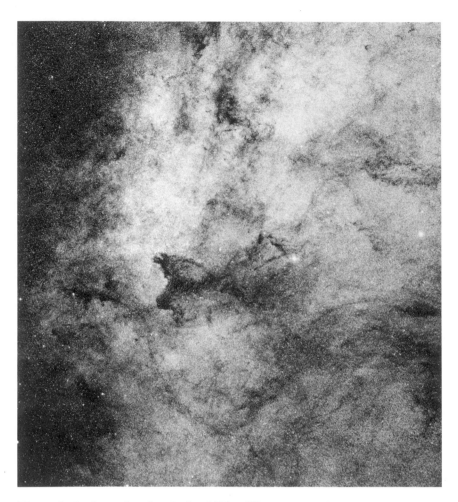

Figure 9–5. *Star clouds of the Milky Way in the constellation Sagittarius with foreground dust clouds.*

indicated by the absorption lines of metals, such as sodium and calcium, that appear in the spectra of early-type stars which normally do not have these lines. These gases do not lie in a cloud surrounding the star in whose spectrum they appear, because their Doppler shift is quite different from that of the other lines in the star's spectrum. Also, we very often see several components of these lines, each having its own radial velocity. Since this occurs especially for more distant stars, we infer that several interstellar gas clouds lie along the line-of-sight between us and the star in each case. There are also large concentrations of neutral atomic hydrogen which emit at a characteristic wavelength of 21 centimeters, in the radio region of the electromagnetic spectrum.

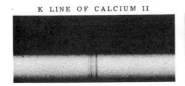
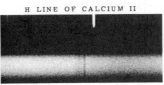
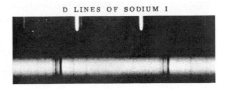

K LINE OF CALCIUM II  H LINE OF CALCIUM II  D LINES OF SODIUM I

Figure 9–6. *Interstellar absorption lines. Absorption lines of Ca II (the calcium H and K lines) and Na I (the sodium D lines) are shown in highly-magnified portions of a spectrogram of ε Orionis. The interstellar lines are very sharp as a result of the very low temperature of the gas (by comparison to the temperature of a typical star). Several sets of lines, each with its own Doppler shift, can be seen, thus indicating that several different interstellar clouds are involved.*

## HII Regions

Of special interest are the hot, glowing clouds of ionized hydrogen that we see around O- and B-type stars. (See color plates 11, 12, and 13.) In the usual notation, neutral hydrogen (that is, with no electron removed, so that it is electrically netural) is abbreviated HI, while ionized hydrogen is abbreviated HII. (This same notation applies to the other elements as well.) These hot hydrogen clouds are often referred to as *HII regions.* (In the older literature they were referred to as *diffuse nebulae,* along with the reflection nebulae.) The gas is largely ionized because the O- and B-type stars, with their relatively high surface temperatures, radiate large amounts of ultraviolet radiation. Photons of this radiation are absorbed by hydrogen atoms; and they have enough energy to ionize those atoms. The more of these stars present, or the higher their surface temperature, the more ultraviolet photons they emit and the larger the amount of gas that the stars' radiation can ionize. It was realized in the 1920s that stars in HII regions are almost always of spectral types O or BO–B5, while stars in the reflection nebulae are of later types (B6 or later). This is reasonable, since stars of spectral types A, F, G, K, and M are not hot enough to emit enough ultraviolet radiation to ionize the hydrogen. As we will see later, the O- and B-type stars and their accompanying HII regions outline the spiral pattern we see in many external galaxies, as well as in our own.

Although it is still not fully understood, there is a strong connection between HII regions and star formation. First, very young clusters (those with O- and B-type stars still on the main sequence) are usually associated with HII regions. Second, in photographs of HII regions we can usually see small dark blobs silhouetted against the glowing gas. (An example, the Rosette Nebula in the constellation Monoceros, is shown in Figure 9–7.) These objects, called *globules,* have been suggested to be in early stages of star formation. Third, there is one case where a star has actually been observed to "turn on" (Figure 9–8), and it occurred in an area of the sky where there are several HII regions and a great deal of dust. Fourth, the observed values of the peculiar velocities of the youngest stars are essentially the same as those of the HII regions and HI clouds.

### Nature of the Interstellar Medium

While astronomers have been interested for three decades in HII regions as a location of star formation, until recently little was known about the physical conditions generally throughout the interstellar medium. At present there is much research into this subject, especially now that ultraviolet and X-ray observations from rockets and satellites give us a much broader range of information. It appears that interstellar space is largely filled with a fairly uniform, low-density gas at a fairly high temperature, possibly as high as $1000°K$. Scattered throughout this "background" gas are dense and relatively cold gas clouds ($10–100°K$) of various sizes. The coldest and densest of these, which have a great deal of dust as well as gas, are "factories" where large molecules, including organic ones like formaldehyde, can be formed and shielded from energetic radiation that would break them up. These low temperatures are inferred from the very narrow widths of the absorption lines due to these molecules, which lie in the radio band of the electromagnetic spectrum. (We remember from Chapter 8 that the thermal broadening is greater as the temperature is higher.)

### Star Formation

The contraction of a gas cloud in free space into a star depends on three factors: the temperature, the mass, and the radius of the cloud. This is because there is a competition between the atoms' tendency to move around on account of their own random thermal motions and the gravitational force exerted by the cloud as a whole. For the cloud to contract, the gravitational force must dominate. This means that the mass of the cloud must be greater than some critical value. If the density is high, the limiting mass value will be smaller. With the typical density

Figure 9–7. *Globules of dust and cool neutral gas in the Rosette Nebula, an HII region. The little dark specks in the close-up are the globules.*

of an interstellar cloud, only stars with considerably larger mass than the sun can form. An ingenious way of getting around this was suggested by Sir Fred Hoyle. A very massive cloud (100–1,000 solar masses) can contract until its density has increased considerably. At that point it can *fragment* into several smaller subclouds, which, though less massive, are sufficiently dense that they can contract. They too may undergo fragmentation until objects less massive than the sun can contract into stars. This process would explain why such a large proportion of stars were apparently formed in larger units like clusters and associations.

The foregoing applies to a cloud in free space. If there is an appreciable pressure on the surface of the cloud, it assists the gravitational force in forcing the cloud to contract. Even if the cloud would not start to contract in free space, the pressure may make it contract. This could occur in or near an HII region (pressure from the hot gas), a supernova

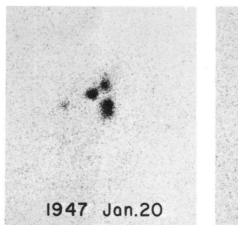
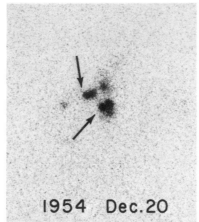

1947 Jan.20       1954 Dec.20

Figure 9–8. *Stars apparently "turning on"— two photographs of the same region taken a few years apart. (The photographs are shown as* negatives *in order to bring out the fine details.) The arrows indicate newly-visible objects in the later photograph.*

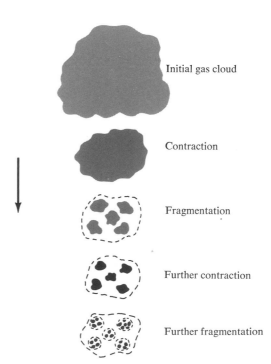

Initial gas cloud

Contraction

Fragmentation

Further contraction

Further fragmentation

Figure 9–9. *Fragmentation of a gas cloud as suggested by Prof. F. Hoyle. A massive initial cloud of gas and dust contracts until the density becomes high enough for small subclouds to "fragment out" and contract on their own. These smaller masses in turn reach a density where still smaller sub-sub-clouds within them can "fragment out" and contract, and so on.*

shell, or with the passage of some other kind of shock wave which causes an increase in the pressure.

## Evolution of Protostars

Once the protostar* has formed, its subsequent evolution depends on its mass. C. Hayashi has shown that all but the most massive stars go through a phase when convection occurs throughout the star's interior. During this phase the protostar decreases in luminosity, with the point

Figure 9–10. *Evolutionary tracks or sequences of points representing different stages in the evolution of a star, for protostars of different masses. The vertical tracks represent contraction according to Hayashi's theory (Hayashi tracks), while the horizontal tracks represent gravitational contraction according to the Kelvin-Helmholtz process (Kelvin-Helmholtz tracks).*

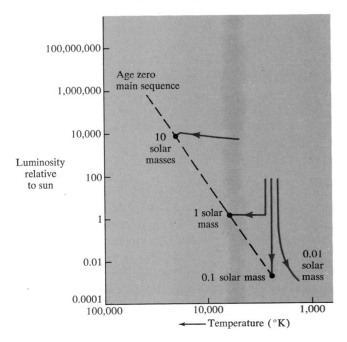

---

* "Protostar" is a term used to describe an object that is becoming a star but has not yet become one. It is considered a star when its nuclear reactions are well-established.

representing the star moving almost straight down in the H-R diagram. The decrease is extremely rapid at first but slows down later. The most massive protostars, on the other hand, have interiors that are in radiative equilibrium. They contract according to the Kelvin-Helmholtz process, with the point representing the star moving almost horizontally across the H-R diagram, until they reach the main sequence. Protostars of intermediate mass first follow a Hayashi evolutionary track (as the vertical descent path is called) in the H-R diagram, then follow a Kelvin-Helmholtz track. In all cases the energy radiated comes from the contraction —the conversion of gravitational potential energy. The nuclear reactions cannot start until the protostar's center has heated up sufficiently, to a few million degrees.

What do protostars look like? In the very earliest stages, when they have not yet contracted to stellar size, they must have very low surface temperatures. In that case, most of their radiation will be at infrared wavelengths. We do see a few infrared sources in regions where star formation is going on (in the constellation Monocerotis, for example, near the Rosette Nebula), and it seems reasonable to suppose that these are protostars. Later, when the protostar has contracted considerably, its brightness fluctuates irregularly. T Tauri variable stars (whose prototype is the star T Tauri) do just this, and they are found only in regions of active star formation. The cause of the variations may be adjustments in the star as the nuclear energy sources are turned on, or it may be obscuration by dust left over after star formation. There are indications that young stars are surrounded by such material.

## The Lives of the Stars

### Limits to a Star's Mass

Once the nuclear reactions have begun and the protostar is located on the main sequence in the H-R diagram, it has become a star. There are, of course, objects that do not become stars. For instance, it can be shown theoretically that if a gas cloud has a mass less than 0.05–0.07 solar masses (the exact value depending on its chemical composition), it cannot heat up sufficiently at its center during contraction to start the nuclear reactions. (Jupiter, with a mass of 0.001 solar masses, falls quite short of this limit.) Such an object will contract until it reaches some limiting size, after which it will remain the same size and simply cool. (We will see the reason for this behavior shortly.) The least massive stars presently known, in the binary system Luyten 726–8, have masses of roughly 0.04 solar masses each. (The mass determination is not so accurate as to prove that they are less massive than the limit given above.) On the other hand, if a star is more massive than 60 solar masses, it will produce so much radiation (remember the mass-luminosity relation!) that it will become unstable—it may even blow itself apart. The most massive star presently known (Plaskett's Star) has a mass of at least 50 solar masses, possibly more; however, no star is definitely known to be more massive than the limit, and Plaskett's Star shows irregular variations in brightness that suggest instability.

### Evolution of Massive Stars

As we saw above, the most massive stars use up their fuel quickly. For a star on the main sequence, the fuel is hydrogen, which is converted by the nuclear reactions into helium. Not all of the star's hydrogen is converted, however. Only the hydrogen in the central regions is converted. When this happens, the nuclear reactions move gradually outwards, much as a forest fire spreads. As this occurs, the outer layers are heated more, so they expand. The star's luminosity remains roughly the same during this *hydrogen shell-burning* phase, while its radius is increasing. By Equation (8–14),

$$L = \sigma T^4 \times 4\pi R^2$$

we see that the surface temperature must decrease. The star thus moves to the upper-right corner of the H-R diagram (Figure 9–11), becoming a bright red giant or, if it is very massive, a red supergiant. With massive stars, this transition from the main sequence to the red-supergiant region is so fast that few stars can be found in between (see Figure 9–2). Thus we have a nearly-empty space in the H-R diagram, called the *Hertzsprung gap* after its discoverer.

While the outer layers have been expanding, the helium-rich innermost regions of the star have been contracting, since no nuclear reactions have been occurring. Of course, as the star's core contracts, it converts gravitational potential energy into heat (thermal energy) and radiation. The heat raises the temperature in the core until the triple-alpha process ignites. Now the star is burning helium to form carbon at the center. The core of the star, when heated up strongly by the helium-burning, may expand enough temporarily to turn off the reaction, but the reaction will finally become well-established. The star becomes both hotter and brighter, and the point representing it moves to the left and upwards in the H-R diagram. At this stage in the star's evolution, the conditions in the star's outer layers are such that they are unstable. The star will pulsate, increasing and decreasing its radius rhythmically. In particular, the star may be a Cepheid variable star (a type of pulsating

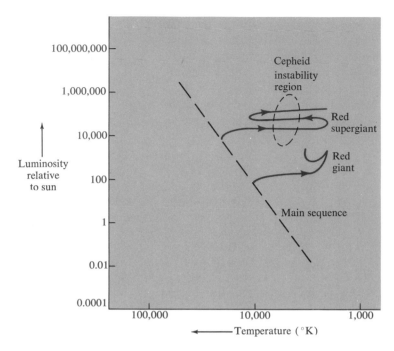

Figure 9–11. *Evolution of massive stars in the Hertzsprung-Russell diagram. The crossing from the main sequence to the red-giant or red-supergiant regions is relatively quick. The region of Cepheid variability is also indicated.*

variable star, mentioned in Chapter 7). Finally the star finds itself on something like a new main sequence, except that helium is being burned instead of hydrogen. After a relatively short while, the helium at the center is used up, and the triple-alpha process moves outwards *(helium shell-burning)*. When this occurs, the point that represents the star in the H-R diagram once more moves horizontally to the right, as the nuclear reactions work their way outwards. Again the star's core contracts and heats up until new nuclear reactions begin at the center. This cycle may be repeated until we reach iron, with the point representing the star crossing the H-R diagram several times. (The later stages of the evolution of massive stars are still somewhat unclear.)

**Supernovae**

We remember from earlier in this chapter that fusion of iron nuclei does not produce any energy. When the star has burned the elements up to iron, it has run out of fuel. At this point there is nothing to keep the star from collapsing. It "implodes" much as a television picture tube does when it is broken (although for a different reason, of course). The core becomes strongly compressed, and the outer layers crash down onto the nearly-rigid core. The outermost layers still have nuclear fuel, and the heat generated by the collapse ignites it explosively. As a result, the outer layers "bounce" off the core and blow into space at a speed of several thousand kilometers a second. Some particles undergo a tremendous acceleration and go off into space as high-energy cosmic rays, and the nuclear reactions generate many gamma-rays (high-energy photons; see Chapter 3) as well.

The process we have just described is believed to account for at least some of the *supernovae*. These stars very suddenly (over a day or so) increase in brightness by hundreds of millions of times. The amount of energy released in a supernova explosion is tremendous, comparable to the lifetime output of a star like the sun. Sometimes the supernovae seen in other galaxies are as bright as, or brighter than, the galaxies in which they appear. The collapse process described above may account for some supernovae. However, several types of supernovae are known, and other processes may well be involved. For example, it has been suggested that objects more massive than the limit of 60 solar masses are so unstable that they explode as supernovae.

If the supernova explosion does not completely destroy the star, a *neutron star* may be left. This object is extremely dense, having several solar masses packed inside a volume with a radius of a few kilometers.

Figure 9–12. *Schematic light curve of a supernova. Note the very dramatic rise in brightness and relatively slow decline.*

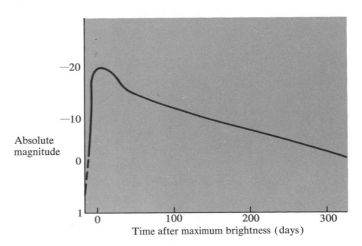

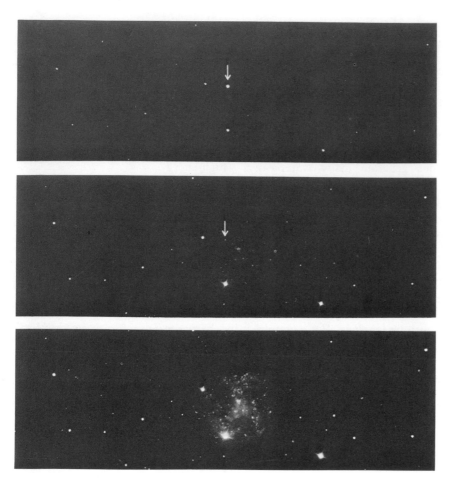

Figure 9–13. *A supernova eruption in a distant galaxy in the constellation Virgo. The supernova is indicated by the arrow.*

At such densities ($10^{14}$ times that of water) atomic nuclei are broken down into individual protons and neutrons. The very high pressure forces the electrons, which normally move around as free particles inside a star, to combine with some of the protons to form more neutrons. The density is so high that the neutrons are touching each other; we say the star is *neutron-degenerate*. Also, the neutron star probably has a very strong magnetic field surrounding it. In 1967, radio astronomers discovered a few sources of extremely regular pulses of radio waves (with periods of the order of a second). Since then more than a hundred *pulsars* (as they are called) have been found, including one inside the Crab Nebula (see Figure 9–15 and color plate 16). This interesting object is the gaseous

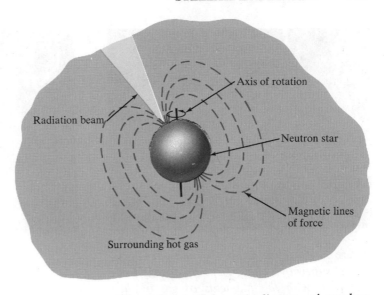

Figure 9–14. *Schematic diagram of a pulsar. The neutron star rotates on its axis, dragging its very strong magnetic field around with it. The magnetic field interacts with the gas surrounding the neutron star to send out a fairly narrow beam of radiation that sweeps around like a searchlight beam as the star rotates.*

shell thrown off in a supernova explosion observed in 1054 A.D. The radio pulses are believed to be produced by gases around the neutron star interacting with its strong magnetic field as the star spins rapidly. The pulsar in the Crab Nebula is located at the same place as the star believed to be the ex-supernova.

The gaseous shell thrown off at high velocity by the supernova is called a *supernova remnant*. The supernova remnant gives off an unusual kind of electromagnetic radiation called *synchrotron radiation*. This kind of radiation is given off by electrons moving at nearly the speed of light in a magnetic field. It was first observed in synchrotrons (nuclear accelerators) and only later considered for astronomical objects. Except in unusual cases, such as that of the Crab Nebula, the energy of the electrons is not high enough for them to give off visible radiation. Normally supernova remnants radiate most strongly at low frequencies in the radio region of the electromagnetic spectrum, except for emission at optical wavelengths. HII regions, on the other hand, radiate more powerfully at high frequencies in the radio region than at low frequencies. In addition to the synchrotron radiation, the remnants give off the

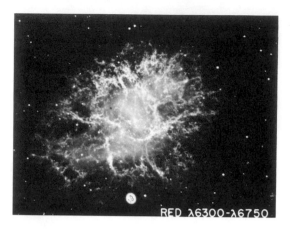

Figure 9–15. *The Crab Nebula, a very young supernova shell. Left: photograph in red light, with the filaments of gas in the shell appearing prominent because of their emission in the red Hα line of hydrogen. Right: photograph in the infrared, showing mainly the continuous-spectrum emission of synchrotron radiation.*

usual emission lines of the heated gases. The remnants expand until they have swept up a considerable mass of interstellar material, after which they gradually slow down and eventually blend into the interstellar medium. An example of a fairly "old" supernova remnant (age about 50,000 years) is the Cygnus Loop (Figure 9–17).

The supernovae are important in the evolution of stars. The supernova remnant sets up an expanding shock wave which may trigger the formation of new stars, as we saw previously. Also, in the supernova explosion, a different kind of nuclear reaction from those we have discussed builds up atoms with higher atomic weights than that of iron. These atoms of the "heavy elements" are mixed with those of lighter elements in the gases of the remnant. Ultimately, they are spread throughout the interstellar medium by the remnant's expansion, to be included in the next generation of stars. The mass of the remnant is not known precisely, but it is at least a significant fraction of a solar mass and may well be as much as 5 or even 10 solar masses.

There is another possible end product for the evolution of a massive star. According to theory, a very massive object will simply collapse—the pressure is not high enough to prevent it. It will become a *black hole*—so dense and compact that electromagnetic radiation cannot escape its extremely strong gravitational field. Such an object would seem to be

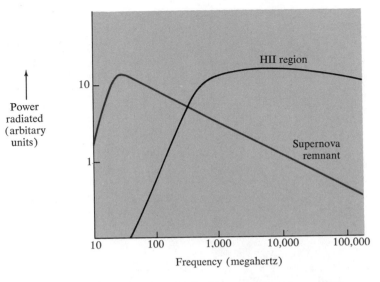

Figure 9–16. *Radio emission from HII regions and supernova shells as a function of frequency. The two kinds of objects are usually quite different in their emission in the radio part of the electromagnetic spectrum. The supernova shells radiate a much larger proportion of their total radio power at low frequencies than do HII regions, as seen in the diagram. (The* hertz *is a unit of frequency, formerly called a* cycle per second.)

undetectable except by its gravitational field. However, in the binary system containing the supergiant star HDE 226868, there is an invisible but fairly massive object giving off X-rays (Cygnus X-1). It has been suggested that Cygnus X-1 is a black hole, with the X-rays coming from hot gases falling in towards the black hole.

## Evolution of Stars of the Sun's Mass

As we saw earlier, only after the massive stars have completed most or all of their evolution do the intermediate-mass stars, like the sun, evolve away from the main sequence in the H-R diagram. We have, of course, an excellent example of a main-sequence star nearby in the sun, which we can study at first hand. We can study such phenomena as the convective motions of the sun's outer layers, the sun's corona, and solar flares—

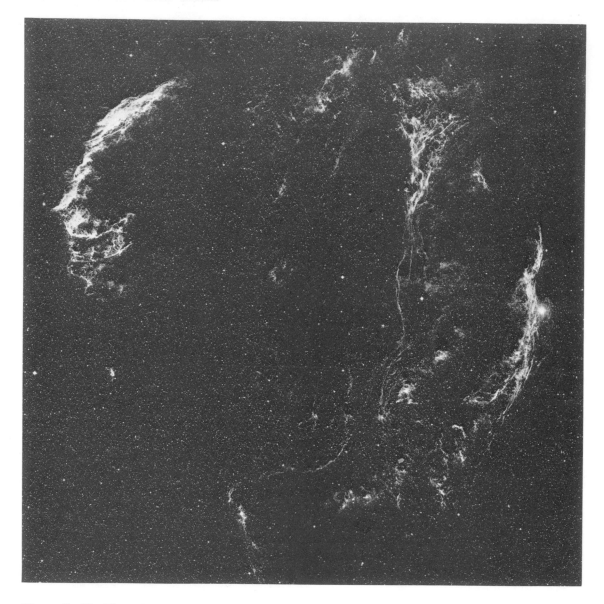

Figure 9–17. *The Cygnus Loop, an old supernova shell, photographed in red light. It has the appearance of a shell, composed of filaments of gas, with one piece missing. No significant amount of continuous-spectrum light is emitted by the Loop in the visible region of the electromagnetic spectrum.*

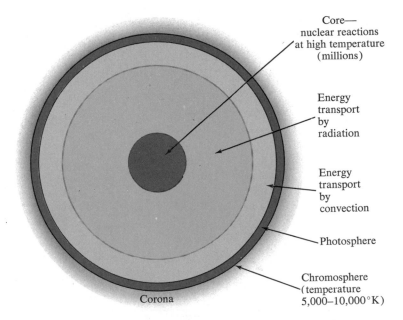

Core—
nuclear reactions
at high temperature
(millions)

Energy
transport
by
radiation

Energy
transport
by
convection

Photosphere

Chromosphere
(temperature
5,000–10,000°K)

Corona

Figure 9–18. *Diagram showing the structure of the sun as determined from observations and from theoretical calculations. The sun is the nearest and thus the best-studied star.*

sudden, intense brightenings of small areas on the sun's surface. At present, according to the age of the earth and theoretical calculations, the sun is some 4–5 billion years old. It thus will be on the main sequence for a comparable length of time before it uses up most of the hydrogen in its innermost regions. Then the nuclear reactions will work their way outwards, and a star of one solar mass like the sun becomes a red giant for the same reason that a more massive star becomes a red supergiant. The point representing such a star moves to the right and also upwards (higher luminosity) in the H-R diagram. The evolution through the red-giant phase is a fairly slow process. When we look at the H-R diagrams of globular clusters, we see large numbers of red giants because they do evolve so slowly. The stars in the clusters are all old, with ages of roughly ten billion years. Only stars less massive than the sun are still on the main sequence. The core of a red-giant star will collapse until it reaches a very high density. At such a high density, the electrons are as tightly packed as they can be, so we say the stellar material is *electron-degenerate.* (The density is much less than that of a neutron star, only a million times that of water.) Electron-degenerate material can be strongly heated without expanding, since the pressure depends mainly on the density of material. Hence, when helium-burning starts, the temperature

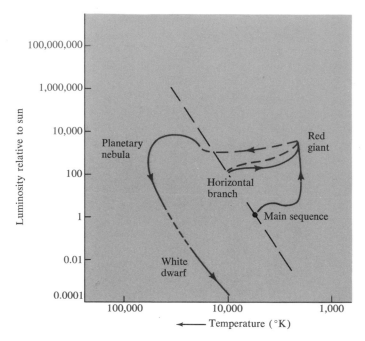

Figure 9–19. *Evolution of a solar-mass star in the Hertzsprung-Russell diagram.*

can rise considerably without the core expanding and cooling off. (Remember, this happens for more massive stars.) The rate at which the triple-alpha process (helium-burning) works increases very sharply as the temperature increases, and as the process speeds up, the temperature rises. As a result, the ignition of helium-burning at the center is sudden and explosive—the so-called "helium flash."

Subsequent evolution is somewhat unclear, but it appears that the point representing the solar-mass star in the H-R diagram then moves to the left. As the helium-burning moves outwards through the star, it evolves towards the red-giant stage once more. The stars in this transition stage are probably the ones occupying the *horizontal branch* in the H-R diagrams of globular clusters (Figure 9–20). Sometime in its later evolution, a star of approximately one solar mass throws off a shell of gas. This is not a dramatic event like a supernova explosion, however. The mass of this shell may be as much as a few tenths of a solar mass, but it has a very low velocity, relatively speaking—a few tens of kilometers per second. What causes the ejection is not certain, but it may be that the radiation pressure in the outer layers during the helium shell-burning phase pushed them off. It has also been suggested that such an instability occurs during the first red-giant phase. In any event, the shell

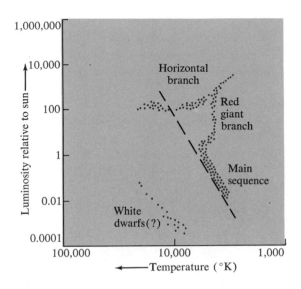

Figure 9–20. *Hertzsprung-Russell diagram for a typical globular cluster (schematic). Note several features: (1) the "turnoff" from the main sequence at a much lower luminosity (and thus lower mass) than in Figure 9–2; (2) the presence of stars in large numbers along the* horizontal branch *and the* red-giant branch; *and (3) a dashed line indicating approximately the limit to the luminosity of stars that can usually be observed. Globular clusters are usually far away, several tens of thousands of parsecs, and stars less luminous than the limit (which varies from one cluster to the next, depending on its distance) will be too faint to be observed and measured.*

expands outwards from the star to form what we call a *planetary nebula.* It has this rather curious name because in the small telescopes of earlier times such objects looked like faint greenish disks, just as the outer planets (Uranus and Neptune) do. A well-known example of a planetary nebula is the Helix Nebula in Aquarius (see Figure 9–21 and color plates 14 and 15). Not all have such a regular appearance, however.

The planetary nebula expands until it dissolves into the interstellar medium. Meanwhile, the star at the center contracts, with its nuclear reactions dying out. As it shrinks, its luminosity remains approximately the same. By our earlier Equation (8–14),

$$L = \sigma T^4 \times 4\pi R^2$$

this means that as the star's radius $R$ decreases, its surface temperature $T$ increases. In fact, the stars at the center of planetary nebulae are some of the hottest stars known, with surface temperatures as high as 200,000°K. Within some 20,000 years after its formation, the nebula has disappeared, and the star has begun to cool. It is now a *white dwarf,* a type of star we mentioned in Chapter 6. The entire star, except for the outermost layers, is electron-degenerate, and it has a density a million times that of water. (The white dwarf was the first superdense type of star to be discovered; its nature has been understood since the 1930s.) All the white dwarf can do now is cool off, changing color from white to yellow to red, and finally becoming invisible. The degenerate star's structure hardly depends at all on the temperature, so it does not change radius. Protostars of very low mass also become electron-degenerate.

Figure 9–21. *The Helix Nebula in the constellation Aquarius (the Water Bearer). This object appears quite faint even in large telescopes. By comparison with other planetary nebulae the Helix has a fairly regular appearance; the majority look quite irregular in shape.*

### Evolution of Low-Mass Stars

The evolution of stars less massive than the sun is not precisely known. We can make theoretical calculations of stellar models, but stars significantly less massive than the sun have not had time to evolve off the main sequence even in the oldest globular clusters. This means we have no way of testing our predictions. However, we expect that their evolution will qualitatively resemble that of stars like the sun, with evolution from the main sequence to the red giant stage, except for the least massive stars. (These stars very likely never start helium-burning because the temperature at the star's center does not get high enough, just as protostars less massive than 0.05 solar masses never start hydrogen-burning.) Presumably these stars also end up as white dwarfs.

If we take the evolutionary track in the H-R diagram for stars with several different initial masses and draw a curve connecting the points on those tracks corresponding to the same age for each star, the curve

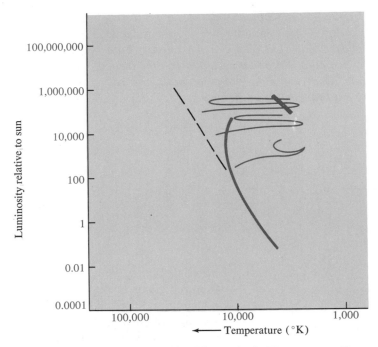

Figure 9–22. *Theoretical Hertzsprung-Russell diagram (schematic). Points corresponding to the same age on the different evolutionary tracks (thin curves) are connected by the heavy curve. Compare with Figure 9–2.*

would represent the shape of the H-R diagram for a cluster of that particular age. In practice, we compare such theoretical curves with the H-R diagrams of real clusters. The turnoff point can give an estimate of a cluster's age, but we can make detailed theoretical calculations to determine both the age and the initial chemical composition of stars in a given cluster. These calculations point to significant differences in chemical composition between old galactic clusters and the globular clusters.

## Ashes to Ashes, Dust to Dust

We have reached the endpoint in our study of the evolution of stars: faint white dwarfs cooling off until they become invisible, tiny neutron stars detectable mainly by their peculiar radiation (radio pulses and X-rays), and possibly (for the most massive stars) black holes. Stars in the course of their evolution lose appreciable amounts of

Figure 9–23. *Comparison of the Hertz-sprung-Russell diagram of an old galactic cluster with that of a globular cluster (both schematic). Note that even though the "turn-off" is at the same point for both, the red giant branches are different.*

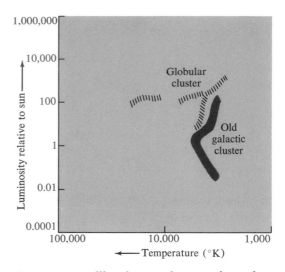

material. It may be a slow process, like the steady mass loss observed from red giants or the relatively sedate ejection of a planetary nebula; or it may be sudden and dramatic, like the explosion of a nova or the more spectacular supernova. All this material ends up as part of the interstellar medium, and some of the material has been "processed" in nuclear reactions in the interiors of stars. As we mentioned before, the gases blown off by supernovae are rich in "heavy elements" (those with high atomic weights). New stars form out of the interstellar gas and dust, so each succeeding generation of stars will have a greater proportion of elements heavier than hydrogen and helium than its predecessor. This gradual increase in the proportion of heavy elements is the process of *enrichment* we mentioned before, and it is a key point of the concept of stellar populations.

## Questions

1. In the conversion from hydrogen to helium (fusion reaction), the fraction 0.007 of the mass is converted into energy. A star typically converts 15 percent of its hydrogen into helium before it evolves away from the main sequence towards the red giant phase. For the sun, the mass is approximately $2 \times 10^{33}$ grams. Use Einstein's equation:

$$E = mc^2$$

$E$ = energy released in ergs
$m$ = mass converted in grams
$c$ = speed of light = $3 \times 10^{10}$ cm/sec

to find out how much nuclear energy the sun has. (Assume that initially the sun consists entirely of hydrogen, which isn't quite true.) For example,

$$E = m \times (3 \times 10^{10})\,(3 \times 10^{10})$$
$$= 9 \times 10^{20}m \text{ ergs}$$

Calculate how much energy the sun will actually use, taking into account the fraction of the sun's mass actually converted into energy. Then, using the fact that the sun radiates away energy at the rate of $1.2 \times 10^{41}$ ergs per year, determine how long it will take for the sun to use up its fuel. (The answer is approximately $10^{10}$ years or 10 billion years.)

2.  If the sun will spend 10 billion years on the main sequence, how long will a star of 10 solar masses remain on the main sequence (approximately)?

3.  Give four observations that indicate the presence of gas and dust in interstellar space.

4.  If the oldest galactic clusters are roughly as old as the globular clusters, how can we explain the fact that the H-R diagrams of old galactic clusters are different from the H-R diagrams of globular clusters? What other factor besides age is involved?

5.  We wish to estimate the effect of "enrichment" on the abundance of helium. Suppose that each generation of stars changes 15 percent of its hydrogen into helium. Then we can set up a table showing the relative numbers of hydrogen and helium atoms in the material from which each new generation of stars is formed.

| | Gen. I | Gen. II | Gen. III | Gen. IV |
|---|---|---|---|---|
| H atoms | 100 | 85.00 | ———— | ———— |
| He atoms | 0 | 3.75 | ———— | ———— |
| fraction | 0 | 0.044 | ———— | ———— |

Of the 100 hydrogen atoms, $100 \times 0.15 = 15$ were converted to helium. Also, four hydrogen atoms are required to make a helium atom, so the 15 hydrogen atoms convert to give $15/4 = 3.75$ helium atoms. As a result, the He/H *abundance ratio* for stars of Generation II is $3.75/85 = 0.044$. Carry out the table through generation IV. Where does the sun, which has a fraction He/H of 0.15, fit into the table?

(As an additional exercise, you might start with 10 percent or 20 percent helium initially and see how the fraction He/H changes over the four generations.)

Note that we assumed that all the star's material is recycled. This is not really the case. Only some fraction gets recycled, with the rest "locked up" in dead stars such as white dwarfs, neutron stars, and maybe black holes.

# 3 GALACTIC ASTRONOMY AND COSMOLOGY

**10** The Milky Way Galaxy

**11** The External Galaxies

**12** Cosmology

# 10 The Milky Way Galaxy

## Introduction

The Milky Way may be seen during clear, moonless nights as a diffuse band of light, describing a great circle on the celestial sphere (Figure 10–1). In the early seventeenth century, Galileo, by means of telescopic observations, was the first to determine that much of the glow could be resolved into a myriad of stars. Later observations with larger telescopes, while revealing more stars than those visible to Galileo, showed that some of the features of the band are not resolvable, and are apparently amorphous masses of glowing gas and dark dust.

From those beginnings, it took over 250 years to fully understand the Milky Way phenomenon in terms of the light originating in a vast disk-shaped system of stars, gas, and dust, of which the sun is a part. Only in the last few decades did it become possible to establish that the Milky Way is a giant spiral galaxy, very similar to our neighboring Andromeda Nebula (Figure 10–2).

Why did it take so long to discover even the grossest features of our galaxy? Part of the answer is obviously the lack of perspective which afflicts us whenever we attempt to study a system of which we are a part. For example, it is not easy to realize the spherical shape of the earth, as long as we remain on it. However, to the astronaut on his way to the moon, the earth will appear obviously spherical. Also, the early difficulty in understanding the relatively simple motions of the planets was due to the lack of perspective. In this case, those simple motions appear to be rather complicated when viewed from the moving platform of the earth. Of course, an observer located just beyond the boundaries of the solar system would have a much more representative view of the system, and he would have never missed, for example, that all the planets revolve around the sun.

Figure 10–1. *A mosaic of the Milky Way composed of several wide-angle photographs. It extends from Sagittarius to Cassiopeia.*

It must be realized, however, that lack of perspective only makes things *appear* different from what they are. The astronomer has long ago learned that in his field things seldom appear to be what they really are. Consequently, he has acquired a healthy mistrust of appearances, and he prefers instead to rely on careful analyses of the observations, where such appearances play only a minor role.

Thus, lack of perspective is only a minor annoyance to the modern astronomer. The long process required to understand the nature of the Milky Way was due to much more fundamental causes. The process of searching for such causes, and of overcoming them as they became understood, demanded the imagination and ingenuity of many people during two and a half centuries. We shall describe the highlights of this process in this chapter.

## Star Counts

William Herschel, in 1785, published the results of the first primitive application of the star count technique. He had selected about 700 regions of equal area distributed randomly in the sky, and in each region counted the number of stars visible through his telescope. He found these numbers to vary greatly from region to region. He then argued that the system of stars of which the sun is a part probably extended farther out in distance in those regions where the star count is

Figure 10–2. *The Andromeda Nebula (M31, NGC 224) is a galaxy very similar to our own. The two compact, fuzzy objects visible in the photograph are small satellite galaxies (NGC 205 and NGC 221).*

high than in those where the star count is low. Putting together all his observations, he concluded that our star system consisted of a flat disk (like a grindstone), with the sun located at its center.

Early in this century, it was known that stars exist having very different intrinsic brightnesses. The brighter stars can be seen at much larger distances than the fainter ones. Thus, the Dutch astronomer J. C. Kapteyn decided that more meaningful results could be obtained if we considered only stars of a given spectral and luminosity class at a time, thus assuring that the intrinsic brightness of all the stars is the same. In this case, when a star (which we will call A) appears to be one magnitude fainter than another star (for instance, B), then A is $\sqrt{2.512} \approx 1.6$ times farther away than B. We are now set not only to obtain a qualitative description of our stellar system, such as was done by Herschel, but hopefully also a quantitative description, based on the reasoning described hereafter.

We now also select many areas in the sky, but without regard to whether they are of the same or of different size. In each region we select the stars of a given spectral and luminosity class, and we count the number of such stars which are brighter than magnitude $m$. Let us call this number $N_m$. Next, let us count in that same region all the stars brighter than magnitude $m + 1$, and call the resulting number $N_{m + 1}$. Since the second count includes all the stars that were counted first, plus some fainter ones, $N_{m + 1}$ is always larger than $N_m$. It turns out that the ratio $N_{m + 1}/N_m$ can tell us a great deal about the distribution of stars around us. Let us first compute the value of $N_{m + 1}/N_m$ if the stars were uniformly distributed in space. For the sake of simplicity, we shall presently assume that the region considered includes the whole sky. In the first count we included all the stars contained within a sphere with the sun at the center and the radius equal to the distance to the faintest star counted. Since the faintest magnitude is $m$, we shall call the distance $r_m$, and the volume of the corresponding sphere $V_m$. In the second count, on the other hand, we included all the stars contained within the sphere whose radius is the distance to the star of magnitude $m + 1$. We call the corresponding distance and volume $r_{m + 1}$ and $V_{m + 1}$, respectively. Now, we saw earlier that

$$r_{m+1} = 1.6 \, r_m$$

Because of the assumed uniform distribution of stars,

$$\frac{N_{m + 1}}{N_m} = \frac{V_{m + 1}}{V_m} = \frac{\frac{4}{3}\pi r_{m + 1}^3}{\frac{4}{3}\pi r_m^3} = \left(\frac{r_{m + 1}}{r_m}\right)^3 = (1.6)^3 \approx 4$$

Now, if this result holds true for the whole sky, it must also hold for any portion of it. Consequently, if in a given region $N_{m + 1}/N_m$ is larger than 4, in that direction in the sky the density of stars is increasing with distance. That would be expected, for example, when you look towards the center of the Galaxy. On the other hand, if $N_{m + 1}/N_m$ is smaller than 4, the stellar population in that direction of the sky is thinning out, and we are obviously looking away from the center of the stellar system.

In Kapteyn's time such counts had been carried out in many regions of the sky, and while the results varied from region to region, most ratios had values between 2.5 and 3.0. Thus Kapteyn concluded that the sun was located at the center of our stellar system. Since the ratios were unequal in different directions of the sky, by assuming that the thinning rate remained constant in each direction he was able to compute for each

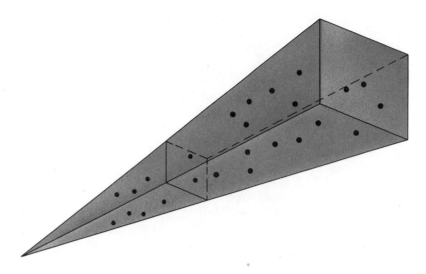

Figure 10–3. *If transparent space were uniformly populated by stars of equal intrinsic brightness, a star of magnitude m+1 would be 1.6 times as far away as a star of magnitude* m. *From this, one expects to find 4 times as many stars brighter than magnitude m+1 as one finds brighter than magnitude* m.

direction the distance at which no more stars existed. In this manner he derived a picture of our stellar system, later known as the "Kapteyn universe," which consisted of an elliptical system of stars having dimensions of 6,000 by 4,000 parsecs, with the sun located at its center (Figure 10–4).

## Cluster Observations

In the years following Kapteyn's work, extensive observational programs were undertaken for the study of galactic (or open) and globular star clusters. As a consequence of those studies, it became clear that the Kapteyn universe could not accurately describe our stellar system. The first significant clue was provided by R. Trumpler's observations of several hundred open clusters. He had felt that the size of the open clusters must be related to the degree of condensation of the stars within them. Thus, by classifying the clusters on the basis of the degree of

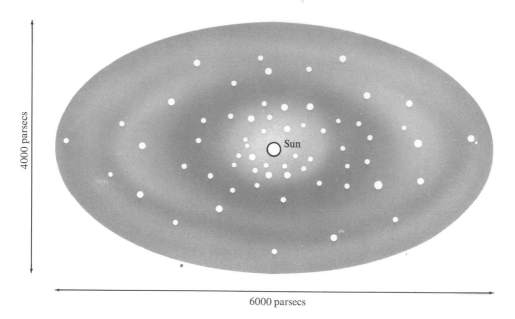

Figure 10–4. *The Kapteyn Universe.*

condensation of their stars, it followed that all the clusters in a given class should have the same size (and also the same number of stars). The apparent angular size that a cluster subtends in the sky is inversely proportional to the distance. Thus if a given cluster subtends about 1/10 the size of another one in the same class, it must be 10 times as far away. In this case, it should be $10^2$ times ($= 5$ magnitudes) fainter. However, he consistently found that the cluster farther away was 6 magnitudes or more fainter. This led Trumpler to think that starlight must be absorbed in part during its travel through the interstellar medium. Here appears the first objection to Kapteyn's work. His prediction that $N_m + {}_1/N_m = 4$ in a uniform stellar system arises from geometrical considerations alone, and is only valid in the absence of interstellar absorption. Kapteyn had been justified in ignoring such absorption, because at the time he did his work nothing had suggested the existence of absorbing interstellar matter. Once this assumption was violated, however, all of his results became highly dubious.

At about that same time, H. Shapley was observing compact globular clusters. By also assuming that globular clusters with a similar degree of concentration were equally bright, he found that the faintest clusters were at distances well beyond the boundaries of the Kapteyn universe! Moreover, the clusters were *not* distributed symmetrically about the sun, but rather about a distant center in the constellation of Sagittarius. He correctly interpreted this to mean that such a point was in reality the galactic center, and that Kapteyn's interpretation of the star counts was erroneous, since the apparent thinning out of stars in all directions was due to the interstellar absorption. Shapley further found that the clusters near the galactic poles became fainter with distance, at a rate closely resembling what one expects from geometry alone, whereas those clusters near the galactic plane become fainter at a higher rate. From this he concluded, again accurately, that the absorption only takes place in a relatively thin layer in the galactic plane.

## Interstellar Absorption

On the basis of the previous results, it became clear that the element complicating the interpretation of the structure of our galaxy is precisely the interstellar absorption. In fact, the absorption is

Figure 10–5. *NGC 4594, the Sombrero, a spiral galaxy in Virgo, seen edge-on. Notice the absorbing band in the plane of the galaxy.*

so severe that along the galactic plane we can see few stars beyond about one-third of our distance to the galactic center. At higher galactic latitudes, on the other hand, we can see not only stars located in the farthest reaches of our galaxy, but also some objects (quasars) all the way to the edge of our visible universe. This non-uniform effect of the interstellar absorption greatly removes the view of our galaxy afforded by the telescope from a truly representative galactic picture. To interpret the meaning of what we see regarding this true picture, we must follow all the observations by very careful analyses. No amount of cleverness, however, allows us to know what is there in those portions of the Galaxy we cannot see. Thus it becomes necessary to search for regions of the electromagnetic spectrum in which the interstellar absorption is so much smaller that we can indeed "see" at least part of what is located in those hidden regions.

For example, a well-known characteristic of interstellar absorption is that it affects the red end of the spectrum much less than the blue end. Consequently, we should be able to see farther in the galactic plane if we look at wavelengths in the far-infrared than if we look in the visual band of the spectrum. This solution has unfortunately met with two obstacles. In the first place, until recently no high quality detectors existed which operated efficiently in the infrared. Only in the last few years has this difficulty been overcome, and infrared astronomy is now an active field of study with a great deal of promise. There is a second difficulty, however, which cannot be eased by technological advances: in general, the brightest objects in the Galaxy emit most of their energy in blue light, with their infrared luminosity orders of magnitude below this maximum. Since faint objects cannot be seen too far away, this *physical* limitation severely curtails the role which infrared astronomy plays in galactic structure studies.

A much more effective way to explore the structure of our galaxy, however, has been found at yet longer wavelengths in the radio region of the spectrum. A major constituent of the interstellar medium is neutral atomic hydrogen. This element emits radiation at a wavelength of 21 centimeters, which is visible throughout the Galaxy. Observations of the 21-centimeter line, described later on in this chapter, have provided the crucial data regarding those regions of the Galaxy we cannot see optically.

## Present View of the Galaxy

### Large-Scale Features

Having isolated the limitations of direct observations for the study of galactic structure, it became necessary to think of ways to circumvent the handicap. For example, Shapley was able to determine the distance to the galactic center from the distribution of the globular clusters. This procedure is still the best one available for that purpose. The current estimate of about 8,000 parsecs (with an uncertainty of about 2,000 parsecs) is very close to the original estimate by Shapley.

While the main body of the Galaxy is confined to a disk, the globular clusters, as we remarked earlier, are distributed spherically around the galactic center. Thus many of them are located well outside the body of the disk. In addition to globular clusters, it has been found that a relatively small number of isolated stars also exist in this larger nondisk region of the Galaxy. Radio astronomical observations, meanwhile, have clearly shown the presence of magnetic fields and of high energy cosmic rays which emit radiation by interacting with such fields. This part of the Galaxy is known as the galactic *halo*. It contains no HII regions or any other indication of either gas plus dust or of early type (that is, young) stars. The diameter of the galactic halo is in excess of 30,000 parsecs. The disk, on the other hand, only occupies a thin slab of the equatorial region of this enormous sphere.

### The Mass of the Galaxy

The best way to determine the mass of any astronomical object is by using Kepler's third law and the Milky Way Galaxy is no exception.

Figure 10–6. *A schematic current view of the Milky Way.*

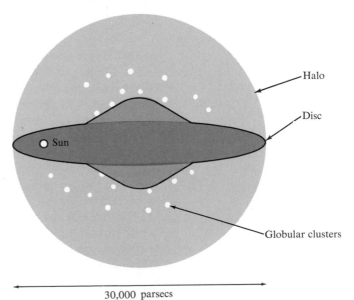

Halo

Disc

Sun

Globular clusters

30,000 parsecs

In this case, we can utilize the orbital motion of the sun around the galactic center, whose characteristics are determinable, even if not directly observable. The sun is the test particle, and we assume that *all* of the galactic mass is contained within the orbit of the sun. Although this assumption is not exactly correct, the error that we introduce by making it is small compared to other unavoidable errors. The ensuing simplification is very great; we can utilize Kepler's third law for the Galaxy, and write

$$M_{\text{galaxy}} = \frac{d^3}{P^2}$$

where $M_{\text{galaxy}}$ is the galactic mass in units of solar masses, $d$ is the distance between the sun and the center of the Galaxy in astronomical units, and $P$ is the period of revolution of the sun around the galactic center in years. Now, while $d$ has already been determined, $P$, usually the simplest quantity to measure, is not directly determinable in this case since we shall see that it exceeds many millions of years and we are not willing to wait that long to find our answers. We can, however, estimate the value of $P$ by noting that the length of the solar orbit in the Galaxy $(= 2\pi d)$ is equal to the galactic orbital velocity of the sun $(= V)$ times the time required for the sun to describe the full orbit which, by definition, is the period $P$. Thus,

$$V \times P = 2\pi d$$

or

$$P = \frac{2\pi d}{V}$$

The problem is now reduced to finding $V_{\text{orb}}$, the orbital velocity of the sun. This can be done, in principle, from the apparent motions of the objects around us which likely do not share in the galactic rotation, such as the globular clusters or even external galaxies. By this procedure, it is found that in one direction in the sky all these objects seem to be approaching us, but in the opposite direction they seem to be moving away from us. When the analysis is completed, we find that $V = 200$–$300$ km/sec in the direction of the constellation Cygnus. Since we expect the galactic orbit of the sun to be nearly circular, the fact that the observed direction of motion is about 90° from the galactic center, as it should be in a circular orbit, makes us feel confident about our measurement.

By using the value of $V = 250$ km/sec, and the currently favored estimate of $d$, we find the period of galactic revolution of the sun to be about 200 million years. A rough estimate of the galactic mass, then, is given by

$$M = \frac{(10,000 \times 200,000)^3}{(200,000,000)^2} = 2 \times 10^{11} \, M_\odot$$

Here, the distance to the center of the Galaxy has been set at 10 Kpc (that is, 10,000 pc), and the conversion rate from parsecs to astronomical units at 200,000 (the actual number is 206,265).

It turns out, then, that the mass of our galaxy equals about two hundred billion times the mass of the sun. Since one solar mass is roughly the average mass of the stars, and since most of the galactic mass is in the form of stars, $2 \times 10^{11}$ is also a good estimate for the number of stars in the Galaxy.

## Detailed Structure

In Chapter 11, we shall see that external galaxies exist in a large variety of shapes. Naturally, we are curious about the exact shape of our own galaxy. We already know that it is disk-shaped, and that it shows enough symmetry to exclude the possibility that it is an *irregular* galaxy such as the Small Magellanic Cloud, one of our close galactic neighbors. We must still determine, though, whether it is a highly flattened elliptical galaxy or a spiral one. All spiral galaxies seen edge on show dark lanes running along the galactic plane, whereas no such bands are visible in elliptical galaxies. Consequently, the existence in our galaxy of strong interstellar obscuration definitely suggests a spiral structure. As usual, however, the ultimate test of this hypothesis rests in the actual observation of the spiral arms. But how can we expect to see spiral arms, despite the handicap of interstellar absorption? To answer this question, we may again examine external spiral galaxies, but this time those which are not seen edge on. In such systems, both the interstellar matter and the most luminous resolved stars are generally confined to the spiral arms. Consequently, we should observe these type of objects in our galaxy when attempting to map its possible spiral arms. Specifically, we should map out O and B stars, and HII regions. By following this procedure, pieces of three spiral arms in the solar neighborhood have been identified. The sun appears to be near the inner edge of one arm, called the Orion arm. The other two arms, one closer to the galactic center (the Sagittarius arm), and the other farther out (the Perseus arm), are located at approximately 2,000 parsecs from the sun (Figure 10–7).

A much more powerful technique for investigating the structure of our galaxy became available when, in 1945, H. C. van de Hulst predicted from theoretical studies that neutral hydrogen should emit radiation at a wavelength of 21 centimeters (in the radio region). We already knew that the spiral arms could be easily mapped if we had a way of mapping

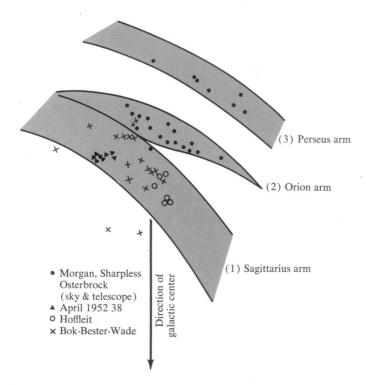

(3) Perseus arm

(2) Orion arm

(1) Sagittarius arm

- Morgan, Sharpless
  Osterbrock
  (sky & telescope)
▲ April 1952 38
○ Hoffleit
× Bok-Bester-Wade

Direction of galactic center

Figure 10–7. *The spiral features of our galaxy as determined by optical means (i.e., observation of O and B stars and HII regions)*

neutral hydrogen, since the interstellar gas is heavily concentrated in the arms, and hydrogen is by far the most abundant gas. H. C. van de Hulst noted that the ground state of the neutral hydrogen atom is formed by two very close *substates* and that a transition from the upper to the lower of the substates would produce a photon of 21 cm radition. Since the 21 cm radiation is not absorbed by interstellar dust, we finally have a means of mapping the spiral arms not only in the vicinity of the sun, but throughout the whole Galaxy. Such work has been carried out, and the results are shown in Figure 10–8. Although the spiral arms are clearly visible, there is no perfect matching between one half of the map and the other half. This is because the 21 cm observations are only interpretable by means of a rotational model of the Galaxy. Such a model is not known perfectly. The observations making up one half of the map were carried out by observers in the Northern Hemisphere and the other half in the Southern Hemisphere. Both groups did not use the identical rotational model, and thus did not obtain perfectly compatible results.

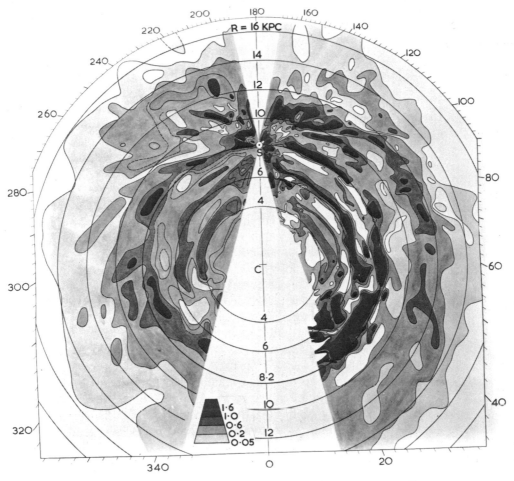

Distribution of neutral hydrogen in the Galaxy (unit=atom/cm³)

based on a model involving both rotation and expansion

Figure 10–8. *The spiral structure of our galaxy as determined from observations of the 21-cm line of neutral hydrogen.*

Basically this means that, while one can indeed trust the numbers of arms detected, their precise location is somewhat in doubt. Also, since the information on which the map is based is the Doppler effect due to the orbital velocity differences of objects at different distances from the galactic center, the maps are incomplete in the regions toward the galactic center and anticenter since there is no radial component of velocity for these two directions.

### Stellar Populations in the Galaxy

W. Baade suggested that the stellar components of our galaxy can be divided in two types: those which belong to the spiral arms, which he called population I objects, and those found in the spaces between the spiral arms, in the central bulge of the galaxy, and in the regions far above and below the galactic plane, which he called population II objects.

Two important observational characteristics differentiate stars of different population type. One is the chemical composition of their surface layers. It is found that population I objects have a much higher metal abundance than population II stars. The second difference is in their general motions; while population I objects always partake of the general galactic rotation, population II objects usually do not. Consequently they may be observationally picked out in radial velocity surveys as "high velocity" objects. This can be understood by visualizing a stalled car in the middle of a heavily traveled highway. If we are in a car moving along with the traffic, all the other moving cars remain at a constant distance from us. However, the stalled car seems to "approach" and "bypass" us very quickly.

In terms of stellar evolution it has now been established that population I stars are young, whereas population II stars are old. This is supported by the location of population I stars in the spiral arms, where the gas and dust (the building material for new stars) are found, as well as by the frequent existence there of short-lived high luminosity objects. Age can also explain the difference in chemical composition, since newer stars are formed from material which is constantly being enriched in heavy elements by supernova explosions.

Finally, all stars born in the early age of the Galaxy, when the Galaxy was presumably poorly organized, will move independently from stars born after a high degree of organization was reached by the Galaxy. This would explain the differences in motion between the two stellar populations.

## Current Problems in Galactic Research

Although we have gone a long way in understanding the structure of our galaxy, the understanding is not complete by any means. Even some of the simplest observations are difficult to explain. For example, it has been determined that stars and gas beyond the central bulge of the Galaxy describe Keplerian orbits around the galactic center. This means that objects farther from the center take a longer time to complete a full orbit around the galaxy than objects closer to it. It is easy to visualize that, in a few turns, the spiral arms should wind up and disappear! However, since the period of revolution of the sun is about 200 million years, and the Galaxy is many billion years old, hundreds of such turns have taken place without destroying the spiral structure. How this has happened constitutes one of the main problems of modern astronomical inquiry. The most accepted current explanation, proposed by C. C. Lin and his co-workers, suggests that spiral arms represent density wave patterns rather than rotating material arms. In other words, stars and gas have their own galactic rotation, except their motion becomes slowed-down once the region occupied by an arm is reached, thus producing higher than average density regions. Since star formation becomes greatly enhanced in such dense regions, new bright stars form there. The amount of light produced in these regions exceeds that produced in the inter-arm regions, by an amount far exceeding the density imbalance.

According to the previous picture, it is not obvious whether those dense regions rotate at all. Even less obvious is whether their rotation would be in the same direction or in the opposite of the rotation of the stars and gas making up the galaxy. In other words, one cannot deduce from this whether spirals precede or trail during the rotation. Recent numerical work has produced galaxies which typically have two armed, trailing spirals. Despite these successes, we still lack a clear understanding of the "triggering mechanism" for the wave patterns. There are cases of barred spirals where spiral arms clearly must be more than mere wave patterns. To this extent at least, the problem of the origin of spiral structure remains very much open.

Research performed during the first half of this century has succeeded in unraveling the process by which a star is formed, determined how its structure is established, and how its evolution takes place. The counterpart of these processes on the galactic scale, however, is not known even in its most rudimentary level. Studying the processes of galactic formation, structure, and evolution must undoubtedly become a major part of the astronomical research effort in the latter part of this century.

## Questions

1. Describe the main difficulties encountered by the astronomer in trying to observe the structure of the Milky Way Galaxy.

2.  If you find 40 stars brighter than magnitude 10 in a given region of the sky, how many stars should you find brighter than magnitude 12 in the same region (if there were no interstellar absorption and the stars were uniformly distributed in that portion of space)?

3.  Two identical star clusters subtend angles of 1 and 10 arc-min respectively.
    (a)   What is the ratio of their distance?
    (b)   What would their magnitude difference be in the absence of interstellar absorption?

4.  Explain what difficulties we would meet in attempting to determine the mass of our galaxy if it only consisted of a population I disk.

5.  How does one map neutral hydrogen in the spiral arms of our galaxy?

# The External Galaxies

## Introduction

Before discussing the subject matter of this chapter, it would be useful to describe the nomenclature of galaxies. Let us, for example, consider the twin galaxy to our Milky Way, which is variously known as the Andromeda Nebula, M31, or NGC 224. First of all, it is remarkable that nothing in any of the three names states unequivocally its nature as a galaxy, rather than, for instance, a planetary nebula or any other of the many types of diffuse objects found in the sky. The reason for this shortcoming is historical, and it originated with the French astronomer Charles Messier during the eighteenth century. Messier's astronomical interest was centered on comets, which can be discovered early on their approach to the sun as *diffuse* (that is, nebulous) faint objects. In order not to be confused by the permanent diffuse objects found in the sky, he decided to catalog the latter as nuisance objects which one should not confuse with comets. His catalog contained about 100 objects, a number determined by his location, and more importantly, the size of his telescope. Messier, of course, did not know he would become much better known by his catalog than by his comet studies. Even today we refer to the most prominent "cloudlike" objects visible from the Northern Hemisphere as M followed by the number in which they appeared in the Messier catalog. Now, since Messier knew nothing about the true nature of the objects in his catalog, it was clearly possible that many different types of objects would be included. This was indeed the case. However, this was not known with any certainty until hundreds of years later, by which time tradition had made it impossible to easily change over to a more descriptive nomenclature.

After that beginning, more complete listings of diffuse objects have been made, including many thousands not seen by Messier. In order to assure completeness, the objects contained in the earlier catalogs were usually also included in the new ones, thus creating several names for the same object. Furthermore, the most striking members of a general class of object acquired their own distinctive names.

The best known catalogs of nebulae are, in chronological order, the Messier Catalog (M), the General Catalog of Nebulae (GC), the New General Catalog (NGC), and two supplements to the NGC called the Index Catalogs (IC). Thus our twin galaxy, which is also prominent enough to have its own name (the Andromeda Nebula), is listed as M31 in the Messier Catalog, and NGC 224 in the New General Catalog.

By about 1910, the number of cataloged nebulae had reached about 15,000, but the nature of only a few of them had been determined. The great majority of them, having somewhat symmetrical shapes, and preferring the regions of the sky far from the galactic plane, were of very uncertain nature. In that regard, two main schools of thought prevailed at that time. On the one hand, many astronomers felt that those nebulae were luminous clouds, probably intermixed with stars, located within our own galaxy. The most prominent proponent of this view was the

Figure 11–1. *M33 (NGC 598), the spiral galaxy in Triangulum, where Hubble observed cepheid variables.*

astronomer H. Shapley. On the other hand, many astronomers, chiefly under the leadership of H. D. Curtis, felt that these nebulae were large aggregates of stars located at distances far exceeding the confines of our galaxy. These views led to the famous Shapley-Curtis debate before the National Academy of Sciences in the year 1920. In that debate, the articulate Shapley was very successful in opposing the view of the extra-galactic nature of these nebulae. In retrospect, that debate was very useful, but not for the reason it was intended. Instead, it showed us that scientific truth cannot be sought in a debate; no amount of brilliant rhetoric can replace hard evidence. The controversy was only truly resolved four years later, when E. Hubble observed cepheid variable stars in M31, M33, and NGC 6822. As we know, cepheid variables obey a period-luminosity relation, so that once the period of variation of the star and its apparent magnitude have been determined, the distance to the star can be found. Hubble's work proved that these nebulae were distant, very large stellar systems—galaxies in their own right.

## Classification of the Galaxies

Galaxies are classified according to a scheme first proposed by E. Hubble. In this scheme, they are divided in four classes: normal spirals (S), barred spirals (SB), ellipticals (E), and irregulars (I).

### Spiral Galaxies

Spiral galaxies are galaxies such as our own and M31. They consist of a disk, with a nucleus and spiral arms, and a corona or halo. Just as in the Milky Way Galaxy, the spiral arms consist of population I objects, while the rest of the galaxy consists of population II stars. In the nearer spiral galaxies, one can see galactic and globular clusters. Dust is often detectable in the galactic plane of these galaxies, especially when we see them edge on.

Spiral galaxies are subclassified into three groups, a, b, or c, according to the degree of openness of the arms, and according to the amount of material in the arms relative to the nucleus. Thus, while at one end of the scale, spiral galaxies consisting of a large nucleus and small, tightly wound arms are classified as Sa, at the other end, spirals having a small nucleus, and wide, loosely wound arms are classified as Sc. Those spirals having intermediate characteristics, such as our galaxy and M31, are classified as Sb.

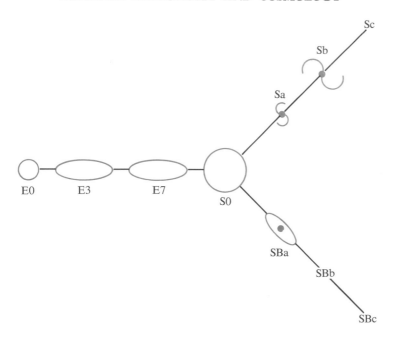

Figure 11–2. *A schematic ("tuning fork") diagram of the Hubble classification of galaxies.*

Figure 11–3. *NGC 5364, an Sc galaxy in Virgo.*

## Barred Spirals

About one fourth of all galaxies displaying spiral structure have bars running through their nuclei, at the ends of which the arms begin. These galaxies are called barred spirals, SB, and they are also subdivided into types, a, b, or c, according to criteria similar to those of the normal spirals. Long exposure photographs recently taken at the Cerro Tololo Interamerican Observatory indicate that the Large Magellanic Cloud (one of our two closest galactic neighbors), which until was thought to be an irregular galaxy, may in fact be a barred spiral (see Figure 11–4).

Figure 11–4. *The Large Magellanic Cloud, as photographed by Dr. V. Blanco at the Cerro Tololo Interamerican Observatory. At the ends of the overexposed bar there are clear indications of faint spiral arms.*

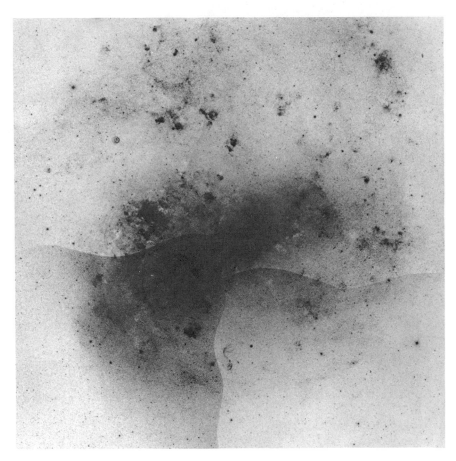

Figure 11–5. *NGC 2523, type SBb, in Camelopardus.*

### Elliptical Galaxies

Elliptical galaxies are galaxies having spherical or ellipsoidal shapes, which consist mainly of population II stars. Elliptical galaxies are found with a great variety of degrees of flattening. This is the feature determining their subclassification in the types E1 (nearly spherical) through E7 (nearly as flat as spiral galaxies seen edge on). There is no detectable dust and only traces of gas within an elliptical galaxy.

### Irregular Galaxies

Irregular galaxies constitute only a small minority of the galaxies in the universe. Their main morphological characteristic is a completely irregular appearance, with no traces of any type of symmetry. The best example of this type of galaxy is the Small Magellanic Cloud. Irregular galaxies contain population I and II objects, with a great profusion of O and B stars, emission nebulae, and interstellar gas, but with little amounts of dust (or none at all).

Figure 11–6. *M82 (NGC 3034), an irregular galaxy in Ursa Major, undergoing a gigantic explosion in its nucleus.*

## The Relative Abundance of Galaxies of Different Types

From a sampling of approximately 600 of the brightest galaxies accessible to Hubble's scrutiny, he concluded that 17 percent of them are elliptical, 2.5 percent irregular, and the rest spiral. More recent studies indicate, however, that these statistics were very strongly biased by the relative brightness of the galaxies of different types. It appears that ellipticals may actually be the most common type of galaxy. In order to produce bias-free statistics, we must first find out how to determine more revealing characteristics of the galaxies than their shapes, such as their mass, luminosity, size, etc., and then determine those properties for a large sample of galaxies. Only after that has been done can meaningful statistics be derived for the galaxies. We are not yet ready to present the results of this formidable task, as it has not been completed. However, we already know how to determine these basic characteristics of the galaxies mentioned earlier, at least for some cases. We shall describe these procedures in the following sections of this chapter. Meanwhile, let us realize that just as in the stellar case, the distance to a galaxy, although by itself not one of its fundamental properties, is the essential first step to determine almost all of these properties.

## Determination of Galactic Distances

The determination of the distances to galaxies is a difficult problem, since their remoteness prevents the use of any direct method such as the trigonometric parallax. Instead, we must rely on the "standard bulb" and the inverse square law method. As explained in earlier chapters, such methods consist of identifying a type of object for which the intrinsic brightness can be found for each member of the group, and determining accurately these intrinsic brightnesses, usually expressed in term of their absolute magnitude $M$. This second step constitutes the *calibration* of the standard bulbs. Since there is effectively no absorption in intergalactic space, it is only necessary to observe the apparent brightness (or apparent magnitude) of one of the standard bulbs to know its distance. As shown earlier, by virtue of the inverse square law for light, we have

$$m - M = 5 \log d - 5 \qquad (11\text{--}1)$$

where $m$ and $M$ are, respectively, the apparent and the absolute magnitude of the object, and $d$ is the distance in units of parsecs. The quantity $m - M$ is referred to as the *distance modulus* of the object.

The most reliable standard bulb used in the determination of distances to galaxies are the cepheid variable stars, because they possess the two main requirements of this task. First, they are bright enough to be seen and resolved at least in the nearest galaxies; second, they adhere strictly to a period-luminosity relation. This tells us what the absolute magnitude of a particular cepheid is just by knowing the period of its variation, a quantity which is very easy to determine. Since $m$ is also easily measurable, $d$ can be readily obtained from Equation (11–1).

Even in this optimum case, however, there is a problem—the absolute magnitudes of the cepheid have been obtained in our own galaxy, by virtue of their membership in clusters to which independent methods of distance determination can be applied. How can we be sure that such calibration is also valid for other galaxies? How do we know that the cepheids in M31, for example, are not systematically brighter, or systematically fainter than those in our galaxy? Each of these possibilities would lead us to underestimate or overestimate their distance, respectively, and hence the distance to M31. This question is not entirely hypothetical, as will be illustrated below.

We know at the present time that there are two classes of cepheids, type I or classical (population I objects), and type II (population II objects). These two types of cepheids have period-luminosity relations which are parallel to each other, yet differing by about 1.5 magnitudes. In other words, a type I cepheid variable of a given period is 1.5 magnitudes brighter than a type II cepheid having the same period. This difference was not known until as late as the early 1950s. The one period-luminosity relation then accepted corresponded rather closely to that of the type II cepheids. On observing an external galaxy not exactly edge-on, on the other hand, one can observe the large majority of the cepheids which are there, and since those of type I are more abundant (and brighter), they make up most of the observed sample. The original distance estimate to M31, then, had been made by looking at mostly type I cepheids, and applying to the observations what was basically a type II cepheids period-luminosity relation. Consequently the absolute magnitude of the objects had been underestimated by about 1.5 magnitudes, and so their distances (and that of M31) had been underestimated by about a factor of 2. Now, the cepheid distance estimates serve in extra-galactic astronomy the same purpose that trigonometric parallax distances serve in stellar astronomy: they form the basis on which all other distance determination methods are calibrated. Consequently, as late as the early 1950s, all extragalactic distance scales were underestimated by a factor of 2. The correction of this error was accomplished by considerations involving another type of variable stars called the RR Lyrae. The most remarkable property of these stars is that they all have an absolute magnitude $M = O$. Those stars, because of their intrinsic faintness, could not be seen in M31 with the 100-inch telescope at Mount Wilson. However, if M31 was at the distance then estimated, these stars should become visible through a 200-inch telescope. Thus, one of W. Baade's first observations with the Palomar 200-inch telescope, completed in 1952, was the search for RR Lyare in M31. When he failed to see them, he correctly concluded that some difficulty existed in the period-luminosity relation for the cepheid variables. Upon investigating the problem further, he discovered the difference in the period-luminosity relation of the two different types of cepheids, corrected the distance to M31, explained why the RR Lyrae had not been seen through the 200-inch telescope, and with one stroke doubled the calculated scale of the universe!

So, as illustrated in this story, we cannot just take for granted that all calibrations performed in our galaxy can be trivially extended to other galaxies. On the other hand, we also saw that obvious errors are usually uncovered by the checks and balances to which all scientific knowledge is subjected. As far as we can tell, after many years of testing, the most general laws that govern nature apply to other galaxies just as well as they apply to our own. The difficulty described above had only appeared because of our lack of understanding the nature of the cepheids in our own galaxy. We are now confident that no future surprises await us in future studies of the cepheid variable stars, and that distance estimates based on their observations are highly reliable.

We have referred to the cepheids as highly luminous stars. This is basically true, since their absolute magnitude $M$ may be as bright as $-6$. On the other hand, they are by no means the brightest objects in the sky: in fact, with the largest telescope now available, they cannot be seen farther than about 6 Mpc (that is, 6 million parsecs). Moreover, since they are extremely rare stars, many nearby small galaxies do not contain any of them. Consequently, the sample of galaxies whose distances have been estimated with the help of the cepheids only consists of about thirty members. For the rest of the galaxies we have to choose brighter objects.

Some blue and red supergiants found in our galaxy are as bright as $M = -9$; novae at maximum light may be as bright as $M = -10$, and finally, supernovae may become as bright as $M = -20$. Because of this, all these objects are used to determine distances to farther galaxies. They have, however, a very basic shortcoming that makes such determinations unreliable—there is a very wide range of possible $M$ values that each of these types of objects may have. As long as we cannot independently determine whether a supernova or a nova is of the brightest, of the least bright, or of a type in between; as long as we cannot fully classify the faint spectrum of distant supergiants to within a very small range of spectral and luminosity classes, we can obtain only very rudimentary estimates of their distances. To that extent, such distance determinations are only useful either in a statistical context, or in reinforcing the reliability of distances determined by other, perhaps equally unreliable, but independent methods.

If galaxies were all equally bright, we could determine their relative distances from their relative apparent magnitudes. Unfortunately, however, galaxies, like stars, may have luminosities anywhere in a range covering several orders of magnitude. For example, elliptical galaxies are found having absolute magnitudes between $-9$ and $-23$, that is, having luminosities between $10^6$ and $10^{11}$ times the luminosity of the sun. Unlike stars, however, it is not easy to tell which ones are the bright ones, and which ones are the faint ones, unless one sees them side by side at a common distance. As we shall see later, galaxies are fortunately often found in large groups called clusters of galaxies, in which they can be seen side by side, relatively speaking. It then appears that the brightest members of different rich clusters of galaxies are reliably of very similar intrinsic lumosities. Thus they can be used to determine the relative distances of the different clusters.

Finally, for the really distant galaxies, there is a distance-determination method whose accuracy is independent of the distance involved. This method is based on the observation, by E. Hubble in 1924, that galaxies recede from us at a rate which is proportional to their distance from our galaxy. While the explanation of this surprising phenomenon is reserved for Chapter 12, it is immediately obvious that, by reversing the argument, one can find the distance to the galaxy from the speed with which it recedes from us. The relation referred to above can be explained quantitatively by means of the Hubble law, which reads

$$V = H \times d \qquad\qquad (11\text{-}2)$$

where $V$ is the velocity of recession, $H$ is the Hubble constant, and $d$ is the distance to the galaxy. $V$ can be readily and accurately measured from the Doppler shift (red shift in this case) of the spectral lines originating in the stars making up the galaxy. Consequently, the most difficult step in this method is the evaluation of $H$, a value which is the same for all galaxies. Indeed, the determination of an accurate value for $H$ has been (and still continues to be) one of the most formidable tasks of extra-galactic research, requiring a large portion of the observing time of the major telescopes. The astronomer mainly responsible for carrying out this task, in the last couple of decades, is A. Sandage of the Hale Observatories. The best current estimates of $H$ are bracketed between 50 and 100 km/sec per Mpc, with the lower boundary more likely to be correct.

## Groupings of Galaxies

At least one half of the more massive stars in our galaxy are found in binary or multiple systems. Even more prominently than stars, however, most galaxies are found in groups consisting anywhere from two members to many thousand members. For example, the Milky Way Galaxy is part of a loose group consisting of at least nineteen members called the *Local Group.* The larger galaxy groupings are called *clusters of galaxies.* The type of groupings in which galaxies are found constitute an important, although not yet understood, clue in the process of galaxy formation. More immediately, it allows us to see many types of galaxies side by side, all located at the same distance from us. This gives us the opportunity to exactly determine not only the relative luminosities of galaxies of all types, but also the ranges of luminosities which galaxies of a given type may have. For example, it has been found that while spiral galaxies may have luminosities between $10^8 L_\odot$ and $10^{10} L_\odot$, and irregular galaxies may have luminosities between $10^7 L_\odot$ and $10^9 L_\odot$, elliptical galaxies may range been $10^6 L_\odot$ and $10^{11} L_\odot$. In other words, both the dimmest and the brightest galaxies found in the universe are of the elliptical type; the former are called *dwarf ellipticals,* and the latter *giant ellipticals.*

This large range of intrinsic brightnesses of the elliptical galaxies makes them very unsuitable as standard bulbs, whenever they are found singly. It appears, however, that a much more reliable standard is constituted by the brightest members of each type found in a rich (that is, heavily populated) cluster of galaxies. Their luminosities seem to be confined within much narrower ranges, and thus they make good standards for measuring relative distances to the clusters. For this reason, distances to galaxies found in rich clusters are much better known than distances to isolated members, or galaxies found in very small groups.

We find the Virgo cluster among the best known rich clusters of galaxies. At a distance of about 10 Mpcs, it is the nearest cluster to us. It occupies about 100 square degrees in the sky (that is, the area of about 400 moons), and it contains more than one thousand members. The next group of this type is located about ten times as far away from us; its name is the Coma cluster. While over 1,000 members of this cluster have been counted, it probably contains fainter members in numbers up to several tens of thousands. To date, many thousands of clusters of galaxies have been identified, largely by the efforts of G. Abell, F. Zwicky, and their coworkers.

Figure 11–7. *The Coma cluster of galaxies. Its mean velocity of recession is V = 1150 km/sec.*

## Radio Studies of Galaxies

The first cosmic radio noise was detected by K. Jansky in 1931, and came from the center of the Milky Way. Since that time, radio emission has been detected from such galactic objects as supernova remnants, HII regions, planetary nebulae, the neutral hydrogen in the spiral arms, and galactic halo. Thus it is clear that from a distance, our galaxy, as a whole, would be detected as a moderately strong radio source. However, the total luminosity of our galaxy in the radio range of frequencies is only of the order of $10^{38}$ ergs/sec. This is a very small fraction of the total luminosity of the galaxy at optical frequencies, which exceeds $10^{43}$ ergs/sec. Under these circumstances (that is, when the optical radiation of a galaxy exceeds the radio radiation by a very large factor), we say that the galaxy is *normal*. Such is the case for the Milky Way. Many examples are known, however, where the radio luminosity of a galaxy is comparable to, or exceeds, its optical luminosity. In such cases, we say that the galaxy in question is a *radio galaxy*. Notable examples of radio galaxies are Cygnus A, NGC 5128, and M87. In most instances, it is found that the structure of the radio-emitting regions in the radio galaxies is complex, often consisting of two large, widely separated components located at either side of the optical galaxy. The spectral distribution (that is, the amount of radiation at each frequency) of the radio radiation originating in these sources is not the one expected from hot gas, but rather resembles the radiation emitted by the halo of our galaxy, which is of the synchrotron type.

Synchrotron radiation is produced by electrons traveling near the speed of light, and spiraling around magnetic field lines.* This suggests the effect of violent events taking place in these objects, an impression that is strengthened by the frequent presence of peculiar features such as jets, which also require a violent origin. It is fair to say that despite a large amount of knowledge gathered in the more than twenty-five years since the discovery of the first radio galaxy, the physical nature of these objects remains a mystery.

## Seyfert Galaxies

Radio galaxies are usually found to be peculiar, mostly irregular in shape when seen through an optical telescope. One almost

---

* The protons are also accelerated by the energetic process that produces the relativistic electrons. However, the relatively large proton mass keeps their velocities well below $c$, and consequently they radiate little energy.

Figure 11–8. *This composite diagram shows the double galaxy from which the Cygnus A radio source originates. The galaxy is in the center of the picture, and the position of the two radio-emitting regions are indicated by the lines.*

expects strange things to be going on in their midsts. However, there is an equally strange type of galaxies which looks remarkably normal to the eye. These are the *Seyfert* galaxies. These galaxies are spirals whose main optical peculiarity is the presence of a very small, very bright nucleus. The surprising feature of the Seyfert galaxies (named after their discoverer C. Seyfert) is that they should perhaps more appropriately be called *infrared* galaxies, since they emit most of their energy in the infrared band of the electromagnetic spectrum. The amount of optical radiation which they emit is that appropriate to a giant spiral (that is, about $10^{43}$ ergs/sec). However, their infrared luminosity, about 100 times larger, places them among the brightest objects in the universe. Since from all the evidence we have they do not appear to be more massive than a typical giant spiral, a Seyfert galaxy must extract on the average, per gram of matter, more than 100 times the amount of energy which is extracted in a normal galaxy. Moreover, since most of the luminosity originates in the nucleus of the Seyferts (which is extremely small), the efficiency of the radiation production per gram of matter in the nucleus is remarkably large. As in the case of the radio

Figure 11–9. *NGC 4151, a Seyfert Galaxy.*
*The bright nucleus appears large because it*
*is overexposed in this photograph in order*
*to reach the rest of the galaxy.*

galaxies, the physical processes underlying the Seyfert phenomenon remain very much a mystery.

## Quasi-Stellar Objects

In the early 1960s, the resolution afforded by radio telescopes became adequate to allow unequivocal identification between radio sources and their optical counterparts in the sky. To everyone's surprise, it was determined that in many instances the site of the radio source appeared as a starlike object in astronomical photographs. While astronomers were prepared to accept the association of extra-galactic radio sources with exploding or colliding galaxies, and radio emission to be produced in such galactic objects as supernova remnants and gaseous nebulae, no one really expected strong radio emission to originate in normal stars. The strongest of the compact radio sources, known as 3C 273 (since it was the radio source number 273 in the Third Cambridge Catalog of Radio Sources), turned out to be a starlike object with a jet. When astronomers at Mt. Palomar (now one of the Hale Observatories) took spectra of the object, they found them not to look like any other stellar spectrum previously known. The puzzle grew more complex when astronomers failed to detect any proper motion of this "star," and concluded that it probably was extra-galactic.

A clue to the extraordinary nature of 3C 273 was obtained in 1966 by M. Schmidt, an astronomer at the California Institute of Technology.

Figure 11–10. *The quasar 3C 273. Notice the jet in the lower right-hand side quadrant. This jet constitutes the stronger component of the radio source.*

He discovered that although the spectral lines in 3C 273 appeared at wavelengths corresponding to lines emitted by no known elements, the ratios of these wavelengths did agree with the ratios corresponding to lines of familiar elements, such as hydrogen.

It is well known that the wavelengths at which spectral lines appear are modified by the Doppler effect, due to an approaching or receding velocity between the light source and the detector on the earth. However, while the wavelengths themselves are changed, their ratios are not modified. Thus Schmidt concluded that the strange appearance of the spectrum of 3C 273 was due to an extreme red shift produced by a velocity of recession between the source (which was first baptized quasi-stellar radio source, then quasi-stellar object, and finally quasar) and the earth. The amount of the red shift corresponded to a velocity of about 15 percent the speed of light. In fact, if one accepted that the object was extra-galactic, and that the red shift originated as a consequence of the general expansion of the universe, then this object was among the farthest objects known! Since its apparent magnitude is about 13, it had to be by far the intrinsically brightest object known, with a visual luminosity of about $10^{45}$ ergs/sec, or nearly 100 times as luminous as our whole galaxy.

Observations later carried out in the infrared compounded the problem by revealing that 3C 273 emitted about 1,000 times as much energy in infrared radiation as it did in the visual range of frequencies.

Figure 11-11. *The photometric history of the brightest quasar, 3C 273, going back to 1888, from a study by Harlan Smith of the University of Texas. The range of brightness fluctuation is about 30 percent. If the quasars are among the most distant objects, then they must be exceedingly bright, even brighter than galaxies, yet the rapidity of their light variations suggests an object very, very much smaller (see text).*

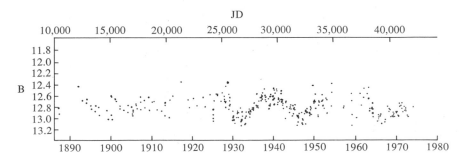

Consequently, we were now dealing with energy outputs amounting to many thousands of times the total energy output from the largest known galaxies. However, the pointlike image of 3C 273 showed that it could not be as large as even a dwarf galaxy.

The next surprise in this strange story appeared when H. Smith and D. Hoffleit of the Yale Observatory, studying old plates of the region of 3C 273 in the Harvard plate collection, determined that this quasar varies in brightness with a period of about 11 years. Now, if any two parts of quasar 3C 273 were further apart along the line of sight than 11 light years, the light produced at a given instant in these different parts would reach the observer more than 11 years apart. In fact, any overall variation of the luminosity of an object would be totally smoothed out, unless the size of the object were small enough so that the light travel time from one end to the other is smaller than the period of the fluctuation. Following this argument, the emitting portion of the quasar 3C 273 must be smaller in diameter than 11 light years (about 4 parsecs). By comparison, our galaxy is larger than about 30,000 parsecs in diameter! Consequently, we are led to conclude that quasars are objects which occupy a volume some thousands of billions of times smaller than that of a galaxy, yet at the same time they emit many thousands of times the energy of a galaxy. This is clearly incredible.

The problem of the quasars has been with us for over ten years. Over 100 quasars have been discovered with red shifts ranging upward to about 90 percent of the speed of light. An unparalleled observational effort has gathered an enormous body of material for most of them. All possible alternatives to the original explanation of the red shifts have been examined. For example, all quasars could be objects ejected at very high speeds from our own galaxy. But then, the total energy required to eject them substantially exceeds the total energy content of our galaxy. Could they perhaps be objects ejected from different galaxies? In that case, some of them should be moving in our direction, and so its spectral lines should be blue shifted. However, we do not see any quasar with a blue shift. All the quasars observed so far have red shifts, and it is totally incredible that this is only due to the effect of chance. Is it possible that the origin of these red shifts is something other than a Doppler effect? After all, light is red shifted when it has to overcome strong gravitational fields in order to leave the object on which it is produced. For example, the light emitted by a white dwarf has been observed to have such an effect. A more extreme manifestation of this phenomenon is expected in the light from a neutron star, although we have not yet been able to make such an observation. However, there are theoretical limitations to the extent of that type of gravitational red shift; and these are far below the largest red shifts observed in the quasars.

Despite the most concentrated effort in the history of astronomy, the physical nature of the quasars is not understood, although many models have been advanced. The chief question that still remains is the origin of the red shift. If the red shift is *cosmological,* that is, if it is related to the Hubble expansion, the enormous distances to those objects (which are by far the farthest objects in the universe) imply that their luminosities exceed those of the brightest known galaxies by factors of many thousands. This fact, coupled with their extremely small size, would appear to indicate that perhaps forms of energy generation unknown to us must exist in nature. This, of course, would not be the first time in which astronomical observations required the existence of unknown forms of energy generation. For example, the energy production in stars remained a mystery until earlier in this century when nuclear energy was discovered. On the other hand, if the red shift is noncosmological, the distances to quasars may be much smaller than now estimated, and their luminosities perhaps much smaller than those of the brightest galaxies. But then what is the mechanism that produces the red shift? Definitely nothing that we already know. It appears that in either case new physics is needed to solve the greatest puzzle of modern astronomy, the nature of the quasars.

## Questions

1. Write down the main types of galaxies found in the universe, and describe their characteristics briefly.

2. How can you tell the difference between an E7 galaxy and a spiral galaxy seen edge on?

3. What is the farthest distance at which we can detect a cepheid variable of $M = -5$ mag, if we can only detect stars with $m < 20$ mag?

4. If $H = 50$ km/sec per Mpc, determine the distance to a galaxy having a red shift corresponding to a velocity of recession of 1,000 km/sec.

5. What is the total luminosity of a galaxy containing $10^{11}$ stars like the sun? Compare this with the total luminosity of a Seyfert galaxy, and of 3C 273, if it is located at the cosmological distance.

6. Explain why new physical processes must be invoked to explain quasars whether they are local (that is, the red shift is noncosmological), or at a distance given by $d = V/H$.

# 12 Cosmology

## Definition

Cosmology is the branch of astronomy whose purpose is to study the organization and structure of the universe as a whole, and by means of this knowledge, trace its past and chart its future evolution. Because of this, cosmology is among the most ambitious and the most exciting of all of man's intellectual undertakings. At the same time, perhaps not surprisingly, it is also one of the most difficult, for the reasons that we shall list hereafter.

The first, and at the same time the most fundamental, difficulty of cosmology is due to the *uniqueness* of the universe. The scientific method is a most efficient tool for discovery when it can be applied inductively—when one can observe many different objects and detect common features that can be explained in terms of a common hypothesis. This, of course, is impossible in cosmology. Consequently, *cosmological models* are proposed not as a consequence of observations, but rather as a consequence of philosophical considerations. Herein lies one of the most basic weaknesses of all cosmological theories. After being compelled to violate the key precept of the scientific method, we should be extra cautious by testing each step with observations if we hope to retain any semblance of scientific reliability. However, even this testing is made nearly impossible by the extreme rarity of observations bearing on this problem. Consequently, for a multitude of years (until the late 1920s), cosmology was a strictly philosophical undertaking, whether it was practiced by Kant, Einstein, Hubble, or anyone else. As of late, however, the situation has changed by virtue of observations which, although few in number, have greatly narrowed down the type of cosmologies that can conform to them.

## The Cosmological Principle

We cannot hope to model the present universe unless we make, a priori, a very fundamental assumption: the portion of the universe accessible to our scrutiny is at least roughly representative of the average conditions in the universe, rather than some peculiar, non-representative sub-region of it. This notion, called the *cosmological principle,* is an often unwritten assumption underlying all cosmological models. Although the requirements of the cosmological principle are formidable indeed, we do not feel overly concerned about its validity, since it appears that we can presently observe a great portion of what we believe to be the whole universe. In that case, what we see cannot fail to represent what the universe is on the average. Of course, if our particular conditions *are* so misleading that the knowledge we currently have regarding the universe is erroneous, then we are left with nothing, not even the hope of ever doing better in the future.

**The Expansion of the Universe**

The first, and one of the most fruitful, cosmological observations was carried out in the late 1920s. It consisted of the monumental work of V. Slipher, M. Humason, and E. Hubble. In simple terms, they discovered that, except for very few nearby galaxies, the light originating in galaxies is always observed to be red shifted when it reaches our telescopes. In fact, whenever the standard distance determination methods are applied, it is found that the amount of red shift is directly proportional to the distance of the galaxy. This result can be written mathematically in terms of the Hubble law, which reads

$$\frac{\lambda - \lambda_o}{\lambda_o} \times c = H \times d \tag{12-1}$$

where $\lambda$ is the observed wavelength of a given spectral line; $\lambda_0$ is the wavelength which the line has at rest, in the laboratory; $c$ is the speed of light; $H$ is the Hubble constant; and $d$ is the distance. It has been mentioned earlier that we know of only two ways of producing a red shift: gravitationally, and by means of the Doppler effect. Since the gravitational fields required to produce the larger observed shifts are much larger than those found in any astronomical object, the very reasonable second assumption is made that the red shifts observed in galaxies are due to the Doppler effect. In that case, the startling conclusion is reached that all galaxies *recede* from us with a velocity proportional to their

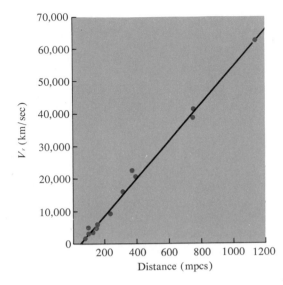

Figure 12–1. *The velocity of recession of galaxies as a function of distance.*

distance from us, or in other words we find that the universe is expanding. The Hubble relation may now more simply be written as

$$V = H \times d \qquad (12\text{–}2)$$

where $V$ is the (radial) velocity determined from the red shift.

There have been several attempts challenging this interpretation. For example, it has been argued that, by an unknown effect, light "tires" on its way to us; that is, it loses energy. Thus it becomes redder. It is fair to say, however, that no serious scheme has yet been proposed to cause this. Consequently, we should not consider this possibility to be any more than something to reconsider if new evidence points to it.

Let us now examine the implications of the "observed" expansion of the universe around us. Since all galaxies recede from us, does it mean that *we* are at the center of something—perhaps at the seat of an explosion which long ago sent all galaxies off on the long voyage they are presently making? From past experience, such a conclusion should make us very suspicious of an error, since we have no evidence whatsoever of any special property of our location. After careful thought, however, we will conclude that the expansion effect observed from our galaxy would be observed equally from any other galaxy in the universe, no matter where the galaxy was located. The best illustration of this point is provided by the analogy of a balloon covered with polka dots, which is being continually inflated. If we could sit on one of the dots (galaxies), we would see all the other dots moving *away* from us as the inflation takes place. If we look at the balloon at a later time, for instance when its size has exactly doubled, we find that, of course, the distance

Figure 12–2. *An expanding balloon resembles the expanding universe. The balloon's surface at (b) is twice as large as it was at (a). The distances between the dots all doubled. Hence, as seen from any dot, the dot farther out moves faster than the closer one.*

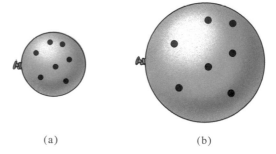

(a)                    (b)

of all the dots from our dot has doubled. This means that during the elapsed time, all dots moved away half their present distance. Therefore, the dots that are far away had to move with a larger relative velocity than those which are closer. Consequently, all the features observed in the receding galaxies are also observed in the motion of the dots on the balloon, no matter which dot we sit on.

## Cosmological Models

Since the early 1930s, it has been known that the universe contains matter visible to us in the form of galaxies, and that those galaxies move away from us with a velocity proportional to their distances. The farthest of the galaxies then observed moved with velocities of a good fraction of the speed of light. With only these facts, there would be a great latitude in thinking up any number of cosmological models, unless additional constraints were accepted. It was then decided that, whenever possible, all known physical laws would be assumed to operate all the way to the scale of the whole universe. It is amazing to realize that under those conditions only two types of cosmologies gained support: the *steady state* cosmology and the *evolutionary* cosmologies.

### The Steady State Cosmology

The basis of the steady state cosmology, whose main proponents are F. Hoyle, H. Bondi, and T. Gold, is the *perfect cosmological principle*. This principle states that what we see is not only representative of the present universe, but also of what the universe has always been and will always be, in a large scale sense. In this scheme, the universe is eternal and infinite, and, in the large scale, always remains the same. In order to retain a constant mean density despite the expansion, however, it was postulated that *continuous creation of matter* must take place. Here we encounter the first violation of one of our cherished natural laws: the conservation of matter (or matter and energy) in the universe. It turns out, however, that the amount of matter creation required is so minute (for example, about 1 atom of hydrogen per 100 years in an average-sized room), that the resulting violation of the conservation of matter law would be totally undetectable by the most accurate measurements that man can perform. Since physical laws are discovered by means of measurements, they are necessarily verified only within those limits defined by the accuracy of the observations. Departures from the law amounting to less than the uncertainty of the measurements could never be detected, and consequently cannot be excluded in principle.

The feature of the concept of continuous creation which is unacceptable to many is that matter is continually being created out of nothing. These same people, however, seem not to be bothered by the thought that matter was created in the remote past, all of a sudden, out of nothing. Of course both of these possibilities are similarly unexplainable in the context of physics, their solution demanding metaphysical considerations which we shall avoid in this book.

## The Evolutionary Cosmologies

Perhaps a more straightforward, although not necessarily simpler, view of the universe contemplates it as consisting of an admixture of matter and radiation in a state of expansion. In such a universe, there are well-known forces operating on the present system, and the present universe arose as a consequence of unknown initial conditions and those same well-known forces acting from the beginning on the system.

Most laws of nature, in their most general form, are much more complicated than the typical versions which are adequate to explain most non-extreme phenomena. For example, Newtonian mechanics is capable of accounting for most problems encountered on earth, and for almost all the features of the planetary motions. If, however, we want to study the motion of objects whose speeds approach the speed of light, Newtonian mechanics is totally inadequate: one must make use of a more sophisticated set of laws called *relativistic mechanics*. Similarly, when one considers relatively small regions of the universe, Euclidean geometry is adequate. On the other hand, when very large portions of the universe are considered, one has to include curvature terms and thus more difficult geometries become applicable. In order to examine the universe as a whole, with its large variety of conditions and its large spatial extent, there is currently only one theory general enough to give realistic results. Such a theory is the *general theory of relativity*. Its mathematical formulation permits the inclusion of almost all types of initial conditions at the beginning of the universe, for example, the effects of any amount of matter or radiation that is present.

A variety of cosmological models can be constructed within this scheme possessing all of the presently observable characteristics of the universe. The only problem of general relativistic cosmologies is that they are so general they do not give useful answers unless all the required initial conditions are fully known. Yet, some of those required quantities are very difficult to measure, and thus we are not sure of the applicability of the final answers. However, at least we know what quantities we need to measure in order to find such answers.

## The Oscillating Universe

Within the framework of general relativity, many models are possible, and many have been proposed as being actually representative of the universe. All of them have in common the fact that the universe is evolving; thus they are called *evolutionary cosmologies*. Of all the versions produced, two general types appear, depending on one of the quantities that is measurable in principle, but not accurately measured yet: the mean density of matter in the universe. If the mean density of matter turns out to be larger than some calculable critical value, the universe as a whole would exert a large attraction on each expanding galaxy, thus eventually bringing the expansion to a halt. Later, this same gravitational attraction will cause a galaxy that was previously moving away from all other galaxies to begin moving towards them; the universe will begin contracting and it will begin heating up. Before the end of this contraction stage, the universe will have become so dense and hot that all vestiges of the preceding stage of expansion will have disappeared. For example, enough energy will have been fed into the heavy elements to convert them into elementary particles from which hydrogen will be made when, in the next expansion, the temperature and density decrease again. Somehow, eventually, the pressure of this highly compact, hot universe is such that it begins reexpanding again.

Such a model would envision the universe to oscillate back and forth between states of high density and temperature, and cold, dilute states. We are, according to this scheme, in the expanding stage of one cycle; however, there is no information regarding the number of previous cycles. In this way, in an oscillating universe, the beginning of it all can be pushed back to a comfortably remote past so as not to bother us.

## The Big-Bang Model

What would happen if the amount of matter in the universe is not large enough to entirely stop the expansion, or, in other words, if the present velocities of the galaxies exceed their escape velocity? This means that the universe will expand forever, decreasing somewhat the velocity of expansion with time. Thus there will be no coming back. It also means that the universe began as a large blob of gas and radiation at enormous temperatures, which started expanding explosively, and such an expansion will go on forever. This cosmological model is known as the *big-bang model,* and it indicates a definite beginning for the universe.

Since the deceleration of the expansion is small in the big-bang theory, we may compute the age of the universe (really an upper limit to it) by the following reasoning. According to the Hubble relation, the velocity with which two galaxies (a distance $d$ apart from each other) move away from each other is

$$V = H \times d$$

Now, if such velocity was the same since the beginning of the universe (at which time the distance between the two galaxies was zero), we have

$$d = V \times t$$

where $t$ is the age of the universe. Thus,

$$t = \frac{d}{V} = \frac{d}{Hd} = \frac{1}{H}$$

In order to apply this formula, it is first necessary to change the units in which $H$ is usually expressed (km/sec Mpc) into (km/sec km = 1/sec).

$$1 \text{ Mpc} = 10^6 \text{ pc} = 3 \times 10^{19} \text{ km}$$

and thus

$$50 \text{ km/sec Mpc} = \frac{5}{3} \times 10^{-18} \text{ 1/sec}$$

$$t = 6 \times 10^{17} \text{ sec} = 2 \times 10^{10} \text{ years}$$

According to this calculation, the universe would be 20 billion years old if the expansion velocity had remained constant. Since there must have been some deceleration, the universe is somewhat younger. In the case of the oscillating universe, this age would only represent an upper limit to the time elapsed since the beginning of the present expansion cycle; but, of course, it would be impossible for us to compute how long ago the universe had its *real* beginning.

*Critical Density and Other Considerations*

Calculations have been made to obtain the "critical density" of the universe, which separates a big-bang from an oscillating cosmology. Its value is $3 \times 10^{-29}$ g/cm$^3$. Meanwhile, a measurement has been made of such mean density considering, of course, only "visible matter," that is, matter that forms galaxies. Such measurement produced a value of $7 \times 10^{-31}$ g/cm$^3$, that is, a value 40 times too small for the universe to be closed. Taken literally, and assuming that no appreciable nonvisible matter exists in the universe, we should conclude that ours is a big-bang (also called *open*) universe. This conclusion would be rather premature, however, since there may be a great deal of invisible matter in the universe, more than enough to close it. This may consist of intergalactic gas, neutrinos,* intergalactic dark objects, and perhaps gravitational radiation. The study is by no means closed.

An alternative way of examining this problem is provided by the following consideration. In view of the finite speed of light, as we look farther and farther away from our galaxy, we are seeing not what the universe is like now, but rather what the universe was like at the time when the light we are seeing initiated its journey through space to come our way. In other words, because of the expansion of the universe, the farther we look, the more dense it should be. This, of course, would not be the case in a steady state cosmology, in which the mean density of the universe remains the same forever. Similarly, if the amount of deceleration of the expansion is small, the density increases only slowly with distance; if the deceleration is larger, it increases more steeply. Now, since the purpose of this measurement is not to obtain a total absolute density, but rather a density change, it is not necessary to include all components of the density (which, as we have seen, is not an easy task), but to take only one—the visible component (galaxies). If the concentration of galaxies (optical or radio) were uniform (as predicted in the steady state), the total number of galaxies seen should go as the cube of the distance. If the concentration of galaxies increases with distance, the curve should turn upward, and this is predicted by all evolutionary models. There is a critical curve which divides the open universe from the closed universe. If the observations' plot turns upwards, but remains below the critical curve, the deceleration is slow; thus the universe is open, and began with a big bang. If the observations, on the other hand, go above the critical curve, then the universe is closed and probably oscillating.

A better known (equivalent) way of showing the influence of evolution in the observations is the red shift-apparent magnitude diagram. Here, we plot the red shift ($\log cz$, where $c$ is the speed of light, and $z = \Delta\lambda/\lambda$) versus the apparent magnitude of the brightest member of rich clusters of galaxies. Figure 12–3 shows several theoretical curves

---

* A neutrino is a particle that has no mass when at rest, and no charge. However, it carries energy away from a very hot gas, or in the course of some nuclear transformations.

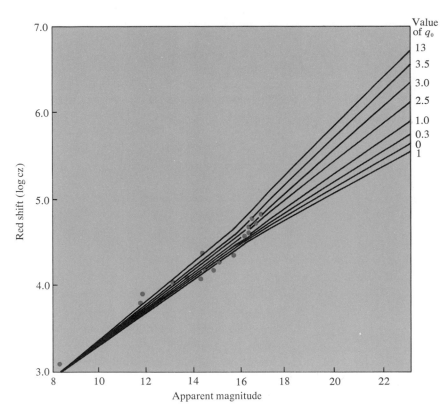

Figure 12–3. *The red-shift—apparent magnitude diagram for galaxies.*

corresponding to different values of the deceleration parameter $q_0$. If $q_0 = 0.5$, the universe would indefinitely remain with a constant size. If the parameter is larger than this, the density of matter in the universe is sufficiently large to halt the expansion and turn it into contraction. Otherwise, the universe will expand forever. Figure 12–4 shows the diameter of the universe as a function of time for different values of $q_0$. We can see from Figure 12–3 that the observations (dots) are still scattered enough so as not to permit us to determine which of the various possibilities is the correct one.

The plots described above are perhaps the most powerful techniques for differentiating among different cosmological models. They are, at the same time, incredibly difficult observations to perform, because the different curves begin departing from each other only at very large distances, which is just where the measurements become unreliable. It is unfortunate that in order to investigate these matters, it is necessary to use the largest telescopes available. As we discussed earlier, there are too few large telescopes and they are subjected to very heavy demands

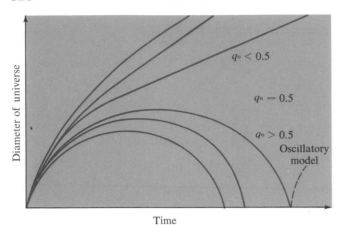

Figure 12–4. *The diameter of the universe as a function of time for different values of the "deceleration parameter," $q_0$.*

for many types of problems. However, we know exactly what to do in order to solve the problem of whether the universe is open or closed, although probably not the best way to do it. Within the next couple of decades, we should have a definite answer.

## The Three-Degree Blackbody Radiation

Cosmological observations are very rare, and when a new one makes its appearance, it causes a great deal of excitement. Most of us were lucky to witness one of these rare events in 1965, when the detection of the three-degree blackbody radiation was announced in the *Astrophysical Journal.*

Two Bell Laboratory scientists, A. Penzias and R. Wilson, were able to obtain in the early 1960s a very low noise antenna that had been built in connection with the early communication satellites. They utilized this antenna to measure the radio noise coming from different directions in the sky at centimeter wavelengths. After many measurements, and after carefully considering all the possible sources of radio noise, there remained a mysterious "three-degree component" whose origin could not be determined.

In 1939, G. Gamow had argued that if our universe is of the evolutionary type, there should remain a vestige of the primeval fireball, that is, the high density and temperature state of the early universe. The vestige would be in the form of equilibrium radiation of a low temperature due to the expansion. In fact, Gamow estimated that a reasonable current value of the temperature of this radiation is about 6°K, a value much too low to be measured at that time.

In 1966, R. Dicke and his coworkers at Princeton concluded that technology had reached the level in which this measurement (6°K) should be feasible. Thus they began building a detector which they thought could do the job. It was at that time that the Bell Laboratory Scientists and the Princeton scientists met to discuss the meaning of the three-degree component. It is not surprising that they soon realized its possible cosmological nature. Also, since the existence of this type of radiation is required in any evolutionary cosmology, but it cannot be easily incorporated into the steady state cosmology, it was clear that the observation favored the former over the latter.

Radiation of that type should have several characteristics. However, when the finding was first announced, these characteristics had not been verified. Such radiation should be totally isotropic (that is, it should have the same value in all directions), and should have the spectral distrubution corresponding to a blackbody at a temperature of 3°K. The very numerous measurements carried out in the intervening years have

Figure 12–5. *The best distribution of the intensity of the primeval blackbody radiation is shown together with the various experimental determinations. The most recent value of the temperature thus determined is actually 2.7°K.*

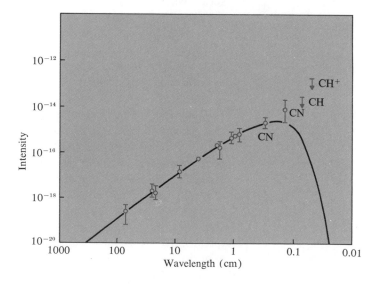

borne out both characteristics. Thus it appears that the observation stands beyond reasonable doubts. We have detected vestiges of the primeval fireball; consequently, the universe is an evolving one. Whether it is open or closed (that is, big-bang or oscillating), however, is a different matter that still remains to be settled.

In closing, we should recall that, in reaching the present picture of the universe and its evolution, important assumptions have to be made, such as the correctness of the cosmological principle, and the Doppler origin of the red shifts observed in remote galaxies. Because of the uniqueness of the cosmological problem, we cannot verify the former assumption, and alternate suggestions have been made to replace the latter. Finally, it is possible to think of yet other cosmological schemes in which all of the few observations are verified. In cosmology, we do not have all the checks and balances which make all other astronomical results much more solidly founded. Unfortunately, because of the nature of the problem, it will always remain uncertain. On the other hand, the clearer the picture becomes, the more our confidence will be strengthened regarding the general accuracy of such a picture.

## Evolution of the Universe

Whether the universe is open or closed does not make a great deal of difference regarding the general sequence of events the universe goes through during its early stage of expansion. Let us describe what such a sequence is according to one of the versions, called the *Einstein-Friedmann universe.*

There is at present no satisfactory theory describing what the universe is like when its radius is almost zero, that is, when the big-bang happens, or when the universe just begins one of its expansions. However, as soon as a millionth of a second has elapsed, the density of matter has reached a value similar to that of the atomic nucleus; the temperature is about $10^{13}°K$. We can then begin to follow the physics involved. At that time no nuclei, not even protons, are stable. They are quickly converted into fundamental particles. The energy density of neutrinos is quite high. By the time the temperature has reached about $10^{10}°K$ essentially all particles and antiparticles have annihilated each other, with the exception of positrons (that is, positively charged electrons). Now we have the matter made up of the more conventional protons, electrons and neutrons, with an admixture of positrons and neutrinos. At temperatures of $10^9°K$ or lower, the positrons have recombined with the great majority of the electrons, and we have the right conditions for element formation. Since the range of temperatures in which elements may form is very narrow, only hydrogen and helium nuclei may form. All of this takes place in less than half an hour. As a result, 70 to 80 percent of the primeval material is hydrogen. All the rest is helium. The heavier elements will only be able to form much later, once stars have been formed, and as a byproduct of stellar evolution.

After the stage of primordial element formation, nothing exciting occurs except that the universe expands and cools down. It is only when the temperature reaches a few hundred degrees that the conditions will allow the condensation of the material into galaxies, and then into stars, and a "live" universe much more familiar to us truly begins.

The evolution of matter after galaxy formation becomes very complicated, since further buildup of the elements begins within the stars, and new energy is fed into the system by the nuclear energy released. However, the primeval matter and the primeval energy parted ways prior to this stage, so that the energy portion does not feel what happens to the matter. Instead, it will retain its blackbody character (that is, the energy is equilibrium radiation totally filling a cavity, which in this case is the whole universe), and will continue to cool down as the universe expands. It is precisely this radiation which has recently been detected, and called the three-degree blackbody radiation, since the current value of the temperature of the primeval radiation is $3°K$.

What will happen after this depends on whether the universe is open or closed. In either case, the universe will continue expanding and cooling down. In the closed universe, however, the expansion will slow down and the universe will reach a maximum radius. Then it will begin contracting, eventually returning to "zero" radius, becoming in the process hotter and hotter. At this stage, the cinders of the previous cycle become totally obliterated. Einstein's theory is unable to predict what follows this stage; however, we believe the universe will re-expand, hence the name oscillatory universe. In the open universe, on the other hand, the expansion will continue forever. After all of the nuclear energy sources are exhausted, the universe will be dead and cold, and still expanding and cooling further, forever.

## Questions

1. Explain why it is necessary to use the concept of the "cosmological principle" in the study of cosmology, a subject which strives to retain its identity as a "science."

2. The "expansion of the universe" is often called a *fact,* indeed one of the few facts really known about cosmology. However, we can

think of ways in which the observations can be explained without expansion. Describe one of these ways and attempt to weigh its merits.

3. Galaxy A is observed to recede from us with a velocity of 3,000 km/sec, whereas galaxy B is observed to have a receding velocity of 9,000 km/sec.
   (a) What is the ratio of their distances?
   (b) What other information do you need to establish the *difference* of their distances?

4. How old is a universe in which H = 600 km/sec Mpc?

5. Argue against the following statement; "The steady state cosmology cannot be correct since it requires continuous creation of matter; hence, it violates the law of conservation of matter and energy."

6. Defend the following statement: "The detection of the three-degree blackbody radiation supports an evolutionary cosmology."

# Appendix 1
## Conversions and Data

### Mass Units

1 gram (gm) = 0.0321 ounce = 0.0022 pound (lb)
$\qquad$ = $10^{-3}$ kilogram (kg)
1 kilogram = 2.205 lb

### Mass Data

Hydrogen atom = $1.673 \times 10^{-24}$ gm
Earth = $5.977 \times 10^{27}$ gm
Sun = $1.991 \times 10^{33}$ gm = 1 solar mass (M)
Galaxy $\simeq 10^{11}$ M

### Length Units

1 angstrom (Å) = $10^{-8}$ centimeter (cm)
1 centimeter = 0.3937 inch (in)
1 kilometer (km) = 0.6214 mile
1 astronomical unit (a.u.) = 149,600,000 km
1 parsec (pc) = 206,265 a.u. (3.26 light years)
1 megaparsec (mpc) = $10^3$ kiloparsecs (kpc) = $10^6$ pc

### Distances

Earth-Moon (ave.) = 384,400 km = 238,900 miles
Earth-Sun (ave.) = 149,600,000 km = 92,960,000 miles
Sun-$\alpha$ Centauri (nearest star) $\approx$ 1.3 pc
Sun-center of Galaxy $\approx$ 8 kpc
Sun-Magellanic Clouds $\approx$ 48 kpc (large), 56 kpc (small)
Sun-Andromeda Galaxy $\approx$ 680 kpc
Radius of visible universe $\approx$ 2 billion pc

## Miscellaneous

Yellow light wavelength $= 5 \times 10^{-5}$ cm $= 5000$ Å

Power output of Sun $= 3.86 \times 10^{33}$ ergs/sec $= 3.86 \times 10^{26}$ watts
$$= 1 \text{ solar luminosity}$$

Power received at Earth's surface $= 1.36 \times 10^6$ ergs/sec/cm²
$$= 1.36 \times 10^{-1} \text{ watt/cm}^2$$

Star density in solar neighborhood $\approx 0.015$ M/pc³

Fraction of Galaxy in hydrogen (neutral) gas $\approx 1\%$

Density of interstellar space $\approx 2 \times 10^{-24}$ gm/cm³
$$\approx 1 \text{ hydrogen atom mass/cm}^3$$

Average density in universe $\approx 10^{-31}$ gm/cm³ ?

Absorption of starlight in galactic plane $\approx 1$ magnitude /kpc (very irreg)

Main sequence lifetime of the Sun $\approx 10^{10}$ years

# Appendix 2
# The Constellations

| Latin (genitive) name | English translation |
| --- | --- |
| Andromeda (Andromedae) | Maiden in chains |
| Antlia (Antliae) | Air pump |
| Apus (Apodis) | Bird of Paradise |
| Aquarius (Aquarii) | Water bearer |
| Aquila (Aquilae) | Eagle |
| Ara (Arae) | Altar |
| Aries (Arietis) | Ram |
| Auriga (Aurigae) | Charioteer |
| Boötes (Boötis) | Herdsman |
| Caelum (Caeli) | Chisel |
| Camelopardus (Camelopardis) | Giraffe |
| Cancer (Cancri) | Crab |
| Canes Venatici (Canum Venaticorum) | Hunting dogs |
| Canis Major (Canis Majoris) | Large dog |
| Canis Minor (Canis Minoris) | Small dog |
| Capricornus (Capricorni) | Sea goat |
| Carina* (Carinae) | Keel of the Argonauts' ship |
| Cassiopeia (Cassiopeiae) | Maiden in a chair |
| Centaurus (Centauri) | Centaur |
| Cephus (Cephei) | King |
| Cetus (Ceti) | Whale |
| Chamaeleon (Chamaeleontis) | Chameleon |
| Circinus (Circini) | Compasses |
| Columba (Columbae) | Dove |
| Coma Berenices (Comae Berenices) | Berenice's hair |
| Corona Australis (Coronae Australis) | Southern crown |
| Corona Borealis (Coronae Borealis) | Northern crown |
| Corvus (Corvi) | Crow |
| Crater (Crateris) | Cup |
| Crux (Crucis) | Cross, southern |
| Cygnus (Cygni) | Swan |
| Delphinus (Delphini) | Porpoise |
| Dorado (Doradus) | Swordfish |
| Draco (Draconis) | Dragon |
| Equuleus (Equulei) | Small horse |
| Eridanus (Eridani) | River Eridanus |
| Fornax (Fornacis) | Furnace |
| Gemini (Geminorum) | The twins |
| Grus (Gruis) | Crane |
| Hercules (Herculis) | Hercules |
| Horologium (Horologii) | Clock |

| Latin (genitive) name | English translation |
| --- | --- |
| Hydra (Hydrae) | Sea serpent |
| Hydrus (Hydri) | Water snake |
| Indus (Indi) | Indian |
| Lacerta (Lacertae) | Lizard |
| Leo (Lenois) | Lion |
| Leo Minor (Leonis Minoris) | Small lion |
| Lepus (Leporis) | Hare |
| Libra (Librae) | Balance |
| Lupus (Lupi) | Wolf |
| Lynx (Lyncis) | Lynx |
| Lyra (Lyrae) | Lyre |
| Mensa (Mensae) | Table Mountain |
| Microscopium (Microscopii) | Microscope |
| Monoceros (Monocerotis) | Unicorn |
| Musca (Muscae) | Fly |
| Norma (Normae) | Carpenter's square |
| Octans (Octantis) | Octant |
| Ophiuchus (Ophiuchi) | Serpent bearer |
| Orion (Orionis) | Orion, the hunter |
| Pavos (Pavonis) | Peacock |
| Pegasus (Pegasi) | Pegasus, winged horse |
| Perseus (Persei) | Perseus |
| Phoenix (Phoenicis) | Phoenix |
| Pictor (Pictoris) | Easel |
| Pisces (Piscium) | Fishes |
| Piscis Austrinus (Piscis Austrini) | Southern fish |
| Puppis* (Puppis) | Stern of the Argonauts' ship |
| Pyxis* (= Malus) (Pyxidis) | Compass on the Argonauts' ship |
| Reticulum (Reticuli) | Net |
| Sagitta (Sagittae) | Arrow |
| Sagittarius (Sagittarii) | Archer |
| Scorpius (Scorpii) | Scorpion |
| Sculptor (Sculptoris) | Sculptor |
| Scutum (Scuti) | Shield |
| Serpens (Serpentis) | Serpent |
| Sextans (Sextantis) | Sextant |
| Taurus (Tauri) | Bull |
| Telescopium (Telescopii) | Telescope |
| Triangulum (Trianguli) | Triangle |
| Triangulum Australe (Trianguli Australis) | Southern triangle |
| Tucana (Tucanae) | Toucan |
| Ursa Major (Ursae Majoris) | Large bear |
| Ursa Minor (Ursae Minoris) | Small bear |
| Vela* (Velorum) | Sail of the Argonauts' ship |
| Virgo (Virginis) | Virgin |
| Volans (Volantis) | Flying fish |
| Vulpecula (Vulpeculae) | Fox |

* The four constellations Carina, Puppis, Pyxis, and Vela originally formed the single constellation, Argo Navis.

# Index

Abell, G., 308
Aberration, chromatic, 95, 98
Above-atmosphere observation, 121–23
Absolute magnitude, 208
  bolometric, 222
Absolute zero, 73, 74
Absorption
  atmospheric, 118
  interstellar, 210, 211, 285–86
Absorption lines, 81
  interstellar, 251, 252, 253
Absorption spectrum, 81
Acceleration, 44–46
  due to gravity, 46–47
Achromatic doublet, 98
Albedo, 151
Alfven, H., 59
Alphonsus (crater), 143
Alpine Valley, 144, 148
Andromeda, Great Galaxy in, 281
  cepheids in, 304, 305
  distance, 304, 305
Ångstrom, 71
Antapex, 214
Aperture, 92
Apex, 213, 214
Aphelion, 49
Apparent magnitude, 27, 112–14
Apollo, 146–49
Associations, stellar, 191
Asteroids, 169
  discovery, 169
  orbits, 169
  sizes, 169

Astroblemes, 146
Astrometry, 192
Astronomical observatories
  optical (table), 99
  radio (see Radio observatories)
Astronomical unit, 36, 202
Atmosphere
  of Jupiter, 165
  lunar, 143
  of Mars, 157
  of Mercury, 151
  of Venus, 154
Atom, 72–73
  absorption and emission of photon, 78
  allowed orbits, 76
  Bohr model, 75
  components, 75
  excited states, 77

Bailey's beads, 21, 25
Becvar Atlas, 112
Big-bang model, 322–23
Binary stars. See Stars, binary
Blackbody, 84
Black holes, 60, 264, 265
Blanco, V., 301
Bohr, N., 75
Bolometric correction, 222
Bolometric magnitude, 222
Bondi, H., 320
Bonner Durchmusterung, 112
Brahe, Tycho, 37
Bright giants, 187
Brightness of an image, 95

Calendar, 16
Carbon cycle (carbon burning), 249
Cassini's Division, 166
Catalogs
  General Catalog, 298
  Index Catalog, 298
  Messier Catalog, 298
  New General Catalog, 298
  Third Cambridge Catalog (3C), 312
Celestrial equator, 7
Celestrial meridian, 8
Celestrial poles
  north, 5
  south, 7
Celestrial sphere, 5
Center of mass, 52
Cepheid, 211, 212, 260
  importance in galactic structure, 304
  types I and II, 304
Ceres, 169
Chromatic aberration. See Aberration, chromatic
Clusters, star. See Star clusters
Color index, 114–15
Color temperautre, 236
Color-magnitude diagram. See Hertzsprung-Russell diagram
Coma cluster, 308
Comets, 170
  Bennett, 172
  coma, 172
  Halley's, 174

Comets (cont.)
  Mrkos, 173
  nature of, 172
  nucleus, 172
  source of, 172
Condensed matter, 59, 261, 264, 267, 269
Conduction, heat, in stars, 250
Constellations, 3
  list of (*see inside cover*)
  seasonal nature, 13
Continuous creation, 320–21
Continuous spectrum, 83
Convection, in stars, 250
Convergent point, 205
Coordinates
  on earth surface, 9
  right ascension and declination, 10
  on sky, 9, 10
Copernican (sun-centered) system, 34
Copernicus, Nicholas, 34
Core
  of Earth, 142
  of Jupiter, 164
  lunar, 149
  of Mercury, 151
  of Venus, 156
Corona
  of galaxy (halo), 299
  solar, 21, 25
Cosmological models, 320–23
Cosmological principle, 318
  perfect, 320
Cosmology, 317ff
  evolutionary, 320, 321–22
  steady-state, 320–21
Crab Nebula, 262, 264
Craters
  lunar, 143
  on Mars, 160
  Mars-moon compared, 160
  on Mercury, 151, 153
  Moscoviense, 147
  on Venus, 134
Curtis, H.D., 299
Curve of growth, 239
Cygnus A, 309, 310
Cygnus Loop, 264, 266
Cygnus X-1, 265

Day
  sidereal, 9
  solar, 9, 11
Deceleration parameter, 325
Declination, 10
Deferent, 32
Density
  of the earth, 142
  of the planets (*table*), 141
Dicke, R., 327
Differentiation, 152
Diffraction, 66
  effects on star image, 119
  formula for, 120
Direct (eastward) motion, 4
Disk, galactic, 279, 281, 287
Dispersion, 95
Distance determination. *See* Parallax
Distance modulus, 209, 304
Doppler effect, 67
  formula, 88
  of galaxies, 318–20
  in radar astronomy, 133
  on starlight, 86
Dwarf star, 187
Dynamical astronomy, 31
Dynamical parallax. *See* Parallax, dynamical

Earth
  age, 176
  as center of universe, 32
  core, 142
  crust, 142
  daily rotation, 5
  density, 142
  drift of plates, 142
  mantle, 142
  origin (*see* Solar system, origin)
  relation to sky, 5ff
  seasons, 13–15
  specifying location on, 9–10
  tides
    in body of, 57, 58
    in oceans of, 57–59
  tilt of axis, 14
  time zones, 12
  yearly motion, 12–13, 16
Eccentricity, 40

Eclipse
  of a satellite by a planet, 48
  of a star by Pluto, 169
  umbral and penumbral shadows, 20
Eclipses, lunar
  appearance, 24
  probability, 23
Eclipses, solar
  annular, 23
  Bailey's beads, 21
  of the near future (*table*), 27
  partial, 20
  probability, 20
  shadow structure, umbra and penumbra, 20
  total, 20
  visibility of corona, 21
Eclipsing binaries, 109
Ecliptic, 15
Eddington, A.E., 59
Effective temperature, 236, 237
Einstein, A., 154, 248
Einstein-Friedmann universe, 328
Electron, 74
  charge, 75
Elements
  abundance, 238
  atomic, 72
Ellipses
  aphelion, 40
  eccentricity, 40
  focus, 40
  perihelion, 40
Emission spectrum, 80
Energy, 63
  absorption by atoms, 75–77
  atomic states, 75–77
  and atoms, 75ff
  carried by waves, 65
  emission by atoms, 77–78
  of photons, 72
  and temperature, 73
  types, 63–64
Energy transport in stars, 250
Epicycle, 32
Equal areas law, in elliptical orbit, 40
Equinox
  autumnal, 16
  vernal, 15

Equinox (cont.)
marking beginning of sidereal day, 10
Equivalent width, 239
Erosion, by micrometeoroids, 150
Eruptive variables, 193
Evening star, 4
Evolution
of low-mass stars, 270
of massive stars, 259
of solar-like stars, 265
Excitation, of atoms, 77
Eyepiece, 93

Falling bodies, 43–44
Finder telescope, 111
Focal length, 93
Focus
Cassegrain, 97
Coudé, 97
of ellipses, 40
Newtonian, 97
Force, 44
of gravity, 46
Frauenhofer, J., 109
Frauenhofer lines, 109
Frequency, 65
Fusion, 248

Galactic (open) clusters. *See* Stars, clusters
Galaxies
barred, 301
classification of, 299ff
clusters of, 307
in Coma, 308
in Virgo, 308
distance determination, 304–7
distribution among types, 303
elliptical, 302
dwarf, 307
giant, 307
energy production, unknown nature of, 310
infrared radiation from, 310
irregular, 302
magnetic fields in, 309
normal, 309
radio study of, 309–10
recession of, 318
Seyfert, 309
spiral, 299

Galaxy, The (Milky Way), 279ff
absorption in, 284
appearance, 183, 279ff
detailed structure, 289–92
halo, 287
mass, 287–89
photomosaic of, 280
rotation of, 280
rotational model, 290, 292
spiral arms, map, 291
21-cm radiation of hydrogen, 289–92
Galilei, Galileo, 41–44
study of motion, 43–44
telescopic observations, 41–43, 279
Gamow G., 326, 327
Gas
ionized, 82–83
neutral hydrogen in The Galaxy, 289–92
as state of matter, 63, 73
Geocentric system, 32
Giants, 187
Globular clusters. *See* Stars, clusters
Globules, 254, 255
Gold, T., 320
Gravitation
acceleration due to, 46–47
formula, 46
tidal force, 56ff
Gravitational collapse, 59–60, 261–63, 264–65, 267
Gravitational contraction, 247
Gravitational redshift, 314
Great Rift, 251
Greenwich, 10
Ground state, of atom, 75

HI, 253
regions, 253–54, 289, 299
Halo, galactic, 287
Hayashi tracks, 257
Heliocentric system, 34
Helium
abundance
in stars, 240–41
in sun, 273
in universe, 73
atomic structure, 76
core burning, 259–60
element, 72–73

Helium (cont.)
flash, 268
fusion reaction, 249
primeval abundance, 329
shell burning, 260
Herschel, W., 280
Hertzsprung gap, 259
Hertzsprung-Russell diagram, 244, 246, 247, 248
turnoff point, 247
Hipparchus, 27
Hoffleit, D., 314
Holmes, S., 185
Horizon, 5, 6
Horizontal branch, 268, 269
Hoyle, F., 256, 320
Hubble, E., 299, 300, 306, 318
Hubble constant, 318
Hubble law, 318, 320, 323
Humason, M., 318
Hydrogen
abundance
in stars, 240–41
in sun, 273
atomic structures, 75–78
core burning, 259
distribution in Galaxy, 289–92
element, 72–73
fusion reaction, 249
primeval abundance, 329
shell burning, 259
Hyperbola, 49

Ideal radiator, 84
Image
blurring, due to atmosphere, 118
of an extended object, 96
formed by telescope, 92ff
of a star, 64
Inertia, 44
Infrared observation, 69, 123–26
radiation from galaxies (*see* Galaxies, infrared radiation from)
Interferometer
radio, 128–32
optical, 230
Interplanetary medium, 140
Interplanetary travel, 51
Interstellar dust, 251
Interstellar gas, 251
Interstellar medium, 251, 254

Interstellar reddening, 115, 211
Ionization of atoms, 82
    dependence on temperature, 82–83

Jansky, K., 126, 309
Johnson, H., 114
Jovian planets, 141
Jupiter, 163ff
    angular size, 164
    density, 163
    energy balance, 163
    Galilean satellites, 165
    Great Red Spot, 164
    *Pioneer 10* photos, 165ff
    rotation, 164
    surface temperature, 165

Kant, I., 317
Kapteyn, J., 281
    universe, 282–83
Kelvin-Helmholtz process, 247
Kelvin-Helmholtz tracks, 257
Kepler, Johannes, 37
    three laws, 40–41
Kirchhoff, G., 109
Kitt Peak National Observatory, 99, 100
Kozyrev, N., 134

La Caille, 36
Lagrange, J. L., 54
Latitude
    effect on view of sky, 5–7
    of a place on earth, 9–10
    relation to declination, 10
Laws of motion, 45
Law
    of gravity, 45
    of reflection, 100
Leap year, 16–17
Lens, 92
    achromatic, 98
Librations (lunar), 149
Light, 63
    diffraction, 66
    Doppler effect, 67
    particle-like properties, 71
    spectrum, 67
    speed, 64
    wavelike properties, 64
        analogy with water waves, 67

Light pollution, 103
Light time, 314
Light wave, 229
Limb darkening, 233, 234, 235
Lin, C. C., 293
Lines
    absorption, 79, 81, 82, 185, 186
    broadening mechanisms, 239
    emission, 78, 79, 81, 186
    in radio region, 290
    shape, 238–40
Lithium, 72
    atomic structure, 76
Lithosphere
    lunar, 150
    Mars, 158
Local Group, 307
Longitude
    as measured on earth's surface, 10
    relation to right ascension, 10
    and time, 12
Luminosity, 181, 189, 221
    class, 187–88
    *table,* 188

M31(NGC 224), 281
M33(NGC 598), 298
M67(NGC 2682), 191
M86(NGC 3034), 303
M87, 309
Magellan, F., 12
Magellanic clouds, 98
    large, 301
    small, 302
Magnetic field
    of Earth, 149, 151
    of galaxies, 309
    of Jupiter, 165
    of Mercury, 151
    of Moon, 149
    of Venus, 156
Magnitude
    absolute, 208ff
        bolometric, 222
    apparent, 27, 112–16
    U.B.V. system, 114–16
    visual, 114
Main sequence, 244
    helium burning, 260

Mantle
    of Earth, 142
    of Mars, 158
    of moon, 158
Many-body problem, 55
Maria, 143
*Mariner 5,* 156
*Mariners 8, 9,* 158–60, 161, 162, 163, 164
*Mariner 10,* 151, 153, 155
Mars, 157ff
    atmosphere, 157
    cratering history, 160
    life on, 163
    mantle, 158
    *Mariner* photos, 157ff
    orbit, found by Kepler, 39
    photomosaic, 159
    polar caps, 161
    rotation, 141
    satellites, 161
    temperature, 157
Mascons, 146, 149
Mass, 44
    of astronomical object, 47
    of satellite, 48
    of stars, 224ff
        limits, 258
Mass-energy equivalence, 248
Mass-luminosity relation, 246
Matter, states of, 73
Mean solar time, 12
Mercury, 150ff
    albedo, 151
    angular size, 150
    atmosphere, 151
    diameter, 151
    magnetic field, 151
    *Mariner 10* photos, 153ff
    rotation, 141, 151
    surface, 151, 153
    in test of general relativity, 154
Messier, C., 197
Messier Catalog, 298
Metals, 73
Meteor "showers," 169, 175
    *table,* 176
Meteorites, 170
Meteoroids, 170
Micrometeorites, 150, 170
Milky Way. *See* Galaxy, The (Milky Way)
Minor planets. *See* Asteroids

Model stellar atmospheres, 237
Model stellar interiors, 249–51
Moon, 3, 142–46
  Alphonsus (crater), 143
  Alpine Valley, 148
  atmosphere, 143
  circumpolar, 143
  density, 149
  eclipses, 23
  far-side, 146, 147
  magnetic field, 148
  manned landings, 146ff
  maria, 143
  Moscviense(crater), 147
  orbit, 20
  origin (see Solar system, for-
    mation)
  phases, 17–20
  seismic study, 150
  Tycho (crater), 148
Moonquakes, 150
Morgan, W. W., 114
Morning star, 4
Moscoviense (crater), 147
Mural quadrant, 38

National Observatory, Kitt Peak,
  99
Neap tides, 58
Nebulae
  diffuse (see HII regions)
  extragalactic, 297ff
  Helix, 270
  Orion, 78
  reflection, 251–52
  Rosette, 255
  solar, 177
Neptune, 168
  discovery, 168
  interior structure, 168
  rotation, 168
  satellites, 168
  surface markings, 168
Neutrinos, 324, 328
Neutron, 74
Neutron degenerate, 262
Neutron stars, 60, 261, 263
Newton, Isaac, 44
NGC 205, 281
NGC 224(M31), 281
NGC 598(M33), 298
NGC 2523, 301
NGC 2682(M67), 191

NGC 3034(M82), 303
NGC 4151, 311
NGC 4594, 285
NGC 5128, 309
NGC 5364, 300
NGC 6205(M13), 192
Nix Olympica, 159
Nuclear reactions in stars, 248,
  249
Nucleus
  atomic, 75
  comet, 172
  galaxy, 299, 310

Objective, 92
Observatories. See Astronomical
  observatories, optical; Radio
  observatories
Oort, J., 173
Orbit
  hyperbolic, 49
  parabolic, 49
  shape, 49
  of space vehicles, 49
  three-body, 54
  two-body, 50
Orbiter (lunar), 146
Orbiting Astronomical Observatory (OAO-2), 121, 122
Orbiting Solar Observatory
  (OSO), 121
Orion, 13
  constellation, 182
  infrared objects in, 124
  Nebula, 78
  photo and map, 182
  spectrum of outer loop, 80

Parabola, 49
Parallax, 199ff
  dynamical 215, 216
  methods of determining
    direct, 199
    indirect, 199, 216
  spectroscopic, 206
  statistical, 212–15
  trigonometric, 199ff
Parsec, 202–3
Penumbral, 20
Penzias, A., 326
Perfect radiator, 84
Perihelion, 49

Period-absolute magnitude
  relation, 211, 212, 213
Photoelectric photometer, 108
Photon, 72
Phototube, 108
Piazzi, G., 169
Pioneer 10, 164, 165
  plaque (see opening drawing,
    Chapter 5)
Planck, M., 84
Planck's law, 84
Planetary nebula, 268–69
Planetary system
  orbital data (table), 140
  origin (see Solar system, for-
    mation)
  overall structure, 139
  physical data (table), 141
  telescopic data (table), 141
Planets, 140ff
  individual properties of (see
    under each planet)
  Jovian, 141
  terrestrial, 141
Plate, photographic, 105
Pleiades, 182
  photo of, 183
Pluto, 168
  discovery, 168
  rotation, 168
Pole star (Polaris), 5
Population I stars, 292
Population II stars, 292
Positrons, 328
Precession, 111
Prime meridian, 10
Production of heavy elements.
  See Supernovae, heavy-ele-
  ment production
Profiles of spectral lines, 237
Proper motion, 203
Proton, 74
Proton-proton chain, 249
Protostars, 257, 269
Proto-sun, 177
Ptolemaeus, Claudius (Pto-
  lemy), 32
Ptolemaic (earth-centered) sys-
  tem, 32
Pulsars, 60, 91, 262
Pulsating variable, 193

Quasars (QSOs), 91, 311–15
   distances to, 314–15
   sizes, 314
   variability, 314

Radial velocity, 87–89, 221
Radiant energy, 64
Radio astronomy, 126, 132
Radio Interferometer, 128–30
Radio observatories, 126–31
   Arecibo, 126
   Bonn, 127
   Culgoora, 131
   Greenbank, 127
   Jodrell Bank, 128
   Parkes, 128
   Westerbork, 131
Radio telescope, 126
   resolution of, 121
Reddening, interstellar, 115, 211
Red shift-apparent magnitude re-
   lation, 324–25
Red supergiant, 244, 259
Reflection effect (in binaries),
   233, 234
Reflection, law of, 100
Reflectors, 95
   Cassegrain, 97
   Coudé, 97
   Newtonian, 97
Reflectron nebulae, 251
Refraction, 92, 93
Refractors, 92, 94, 97
Relativity, 154, 321
Resolution, 119, 119–21
   of a radio telescope, 121, 128–
   31
Retrograde motion, 4
Retro-reflectors, lunar, 149
Revolution. See also under indi-
   vidual objects
   periods of planetary, about sun
   (table), 140
   of sun about center of Galaxy,
   288
Right ascension, 10
   origin of, 10
Rills, lunar, 143
Roche, E. A., 57
Rotation. See also under individ-
   ual objects
   and broadening of stellar lines,
   239

period of planetary, on axis
   (table), 140
planetary, deduced by radar,
   132–33
of spiral arms in a galaxy, 293
RR Lyrae stars, 211, 305
Russell, H. N., 187

Sagittarius, 107, 284
Sandage, A., 307
Satellites of planets, origin, 177
Saturn
   Cassini's division (photo), 166
   rings, 166
   rotation, 166
   satellites, 166
   surface temperature, 166
Scalar, 45
Schmidt, M., 312
Scientific revolution, 31
Seasons, 13–15
Seeing, 118, 119
Selection effects, 194
Semimajor axis, 225
Seyfert galaxies, 309–11
Shadows, umbral and penum-
   bral, 20
Shapley, H., 284, 299
Sidereal day, 9
Sky
   apparent daily motion, 5
   celestial sphere, 5
   as a clock, 7
Sky charts, 110
Skylab, 91
Sky Survey National Geographic
   Society-Palomar Observa-
   tory, 107–12
Slipher, V., 318
Smith, H., 314
Solar day, 11–12
Solar motion, 212
Solar nebula, 177
Solar system, 176ff
   age, 176
   Copernican scale model, 34
   formation, 176
Solar wind, 173
Solstices, 16
Spectral classification, 184ff
Spectral types, 184
   special, 189
   table, 187

Spectograph, 109
   high-dispersion, 110
Spectroscopic binary, 193
   double-lined, 193
   single-lined, 193
Spectroscopic parallax. See Parallax, spectroscopic
Spectroscopy, 109–10
Spectrum
   absorption, 80
   continuous, 83
   emission, 80
   of a star, 64
Spectrum binary. See under individual binaries
Spiral arms, 289ff
   direction of rotation, 293
   map of, 291
   noted in Large Magellanic
   Cloud, 301
   triggering mechanism, 293
Spiral density waves, 293
Spring tides, 58
Star clusters
   age determination, 244
   chemical composition differences, 271
   globular, 191, 272
   open, 191, 272
Star counts, 280ff
Stars
   age determination, 244
   angular diameters, 228–29
   birth, 251, 254, 256, 259
   chemical composition, 181,
   240–41
   circumpolar
   north, 6
   south, 7
   clusters of (see Star clusters)
   collapsed, 59, 261, 264, 267,
   269
   distances (see Parallax)
   energy sources, 247ff
   evolution, 259ff
   high-velocity, 292
   lifetime, 246
   luminosities, 189ff
   magnitudes (see Magnitude)
   masses (see Stars, binary)
   model atmospheres, 237
   model interiors, 249–51

Stars (cont.)
  population I, 192, 292
  population II, 192, 292
  radii, 229–31, 232
  surface temperatures, 181, 235–39
  variable, 108, 193, 211, 260
stars, binary, 190ff
  astrometric, 192, 193
  eclipsing, 194, 195, 196, 229–35
  histogram of separations, 196
  mass determination, 224–25, 226–28
  radii, 229–31
  semimajor axis, 225–26
  spectroscopic, 193, 194, 222, 227
  spectrum, 193
  visual, 51, 192, 195ff
Stars, individual
  Alcor, 54
  BD+73°8031, 189
  Cygnus X-1, 265
  δ Cephei, 211
  ε Orionis, 253
  η Hercules, 188
  40 Eridani, 53
  HDE 226868, 265
  Krüger 60B, 184
  Mizar (ζ UMa), 54, 194
  Nova Persei 1901, 207
  RR Lyrae, 211, 305
  Sirius (αCMa), 193
  T Tauri, 258
  13 Ceti, 53
  Vega (αLyrae), 213
  WW Cygni, 195
States of matter, 73
Statistical parallax. See Parallax, statistical
Steady state cosmology, 320
Subgiants, 187
Sun, 265, 266
  spectral type, 190
Supergiants, 187
Supernovae, 260–65
  heavy-element production, 292
  Tycho's, 37
Surface gravity. See Luminosity class

Surveyor, 146
Synchrotron radiation, 309

Tangential velocity, 203
Telescope, optical, 92ff
  as an analytical instrument, 109
  as a camera, 105
  focal length, 93
  as a light meter, 107
  magnification, 93
  reflecting, 95ff
  refracting, 92ff
  resolution, 119–21
  Schmidt, 107
  sites, 100–105
  use of, 117
  world's larger (table), 99
Telescope, radio. See Radio telescope
Temperature
  of background radiation, 326–28
  and blackbody radiation, 84–85
  color, 236
  effective, 236–37
  and excitation, 76, 237
  of incandescent objects, 84–85
  and ionization, 82–83, 257
  of main sequence stars (table), 238
  moon, 143
  nature of, 73
  planets, surface
    Jupiter, 165
    Mars, 158
    Mercury, 151
    Neptune, 168
    Saturn, 166
    Uranus, 167
    Venus, 154–55
  scales of, 73
  stars, outer layers, 235ff
  sun, 84
  universe, at early epoch, 328
Terrestrial planets, 141
Thermal broadening, 239
3C 273, 312–14
3C 295, 103
3°K blackbody radiation, 326–328

Tidal forces, 55–59, 232–33
  in binary stars, 232–33
  on body of earth, 57
  on oceans, 57
Time
  mean solar, 12
  sidereal, 9
  solar, 9
  zones, 12
Triangulation. See Parallax, trigonometric
Trigonometric parallax. See Parallax, trigonometric
Trumpler, R., 283
Turnoff point, 247
21-cm line, of hydrogen, 290
Two-color diagram, 115
Tycho (crater), 145

U,B,V-system, 114
UHURU, 122, 123
Ultraviolet radiation, 65, 70
  atmospheric opacity, 118
  studied via balloons and rockets, 121
  studied via satellites, 122
  ultraviolet magnitude, 114
Umbra, 22, 23
Units and conversions. See appendix
Universe
  age, 323
  critical density, 323, 324
  deceleration, 324
  diameter, 326, 328
  Einstein-Friedmann, 328
  evolution, 328–29
  expansion, 318–20
  oscillating, 322
  uniqueness, 317
Uranus, 167
  angular size, 167
  density, 141
  discovery, 167
  interior, 167
  rotation, 167
  satellites, 167
  surface temperature, 167

Van de Hulst, H. C., 289
Vector, 45

*Venera,* Soviet Venus probes, 155

Venus
atmosphere, 154ff
diameter, 155
magnetic field, 156
radar image, 134
radar studies, 155ff
rotation, 156
surface temperature, 164–65

Vernal equinox, 15
Very Large Array (VLA), 131
Virgo cluster, 308
Visual binaries. *See* Binaries, visual
Visual magnitude, 114

Wavelength, 65
Wave motion, 64ff
White dwarfs, 60, 189, 269

Width of spectral lines, 238, 239
Wilson, R., 326
Windows, atmospheric, 123
Wolf-Rayet stars, 189, 190

Year, length, 16

Zenith, 7
Zwicky, F., 308

# North Polar Chart

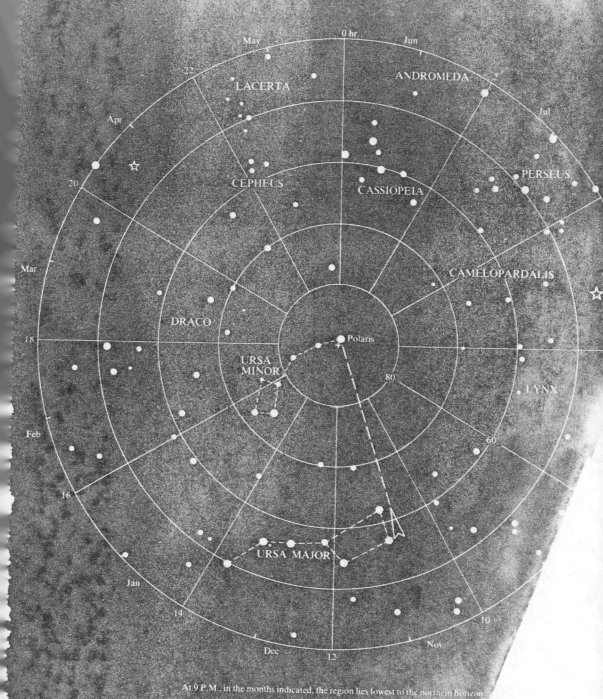

At 9 P.M., in the months indicated, the region lies lowest to the northern horizon

**Face north and find the North Star (Polaris) using the B... "pointer" stars (arrows). Time and date are as in the Equ... Chart, inside front cover.**

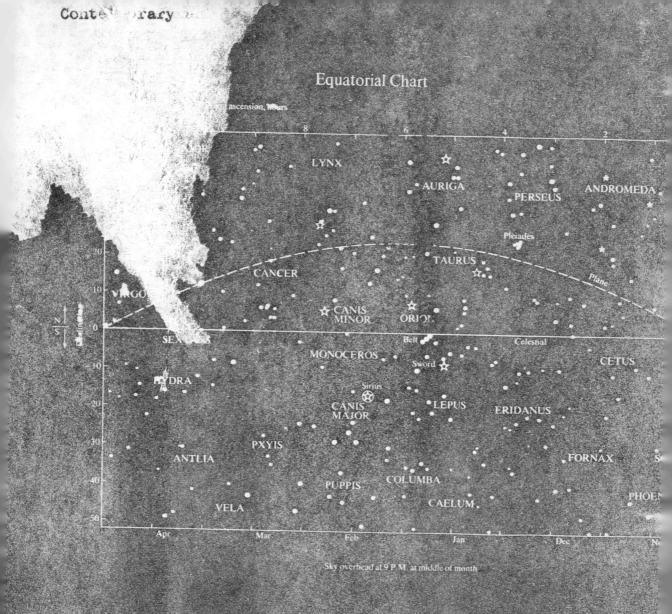

## Equatorial Chart

Sky overhead at 9 P.M. at middle of month

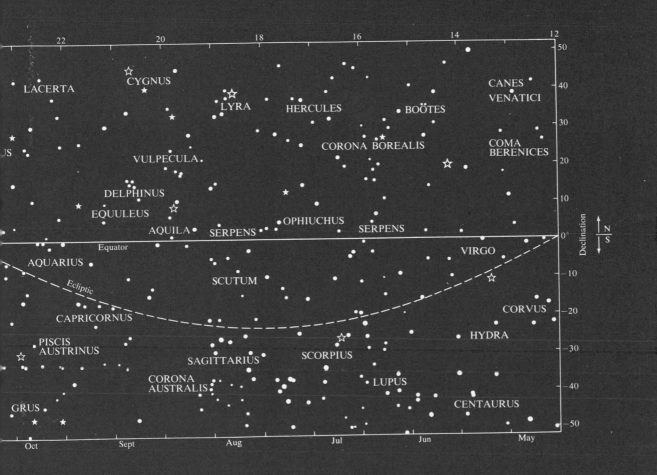

This chart shows the stars as seen by an observer facing south and looking up at the sky. The dates at bottom indicate the region then crossing
the meridian at the time on the chart. For each hour later (earlier), shift
left (right) an hour of right ascension. For scale, note that the moon fills
½° on the declination scale.

8-22-75